FAMOUS AIRCRAFT OF THE
NATIONAL AIR AND SPACE MUSEUM

VOLUME 1
EXCALIBUR III
The Story of a P-51 Mustang

VOLUME 2
THE AERONCA C-2
The Story of the Flying Bathtub

VOLUME 3
THE P-80 SHOOTING STAR
Evolution of a Jet Fighter

VOLUME 4
ALBATROS D. Va
German Fighter of World War I

VOLUME 5
BLERIOT XI
The Story of a Classic Aircraft

Blériot XI
The Story of a Classic Aircraft
by Tom D. Crouch

PUBLISHED FOR THE

National Air and Space Museum

BY THE

Smithsonian Institution Press

WASHINGTON, D.C.

1982

*For Louis Blériot, who built the aircraft; John Domenjoz,
who flew it; and Paul Edward Garber, who preserved it.*

LIBRARY OF CONGRESS CATALOGING IN PUBLICATION DATA

Crouch, Tom D.
  Blériot XI, the story of a classic aircraft.

  (Famous aircraft of the National Air and Space Museum; v. 5)
  Bibliography: p.
  1. Blériot XI (Aircraft) I. National Air and Space Museum. II. Title. III. Series.
TL686.B6C76   629.133'343   81-607931
ISBN 0-87474-345-1   AACR2

Cover art by John F. Amendola, Jr.

*Frontispiece: The Blériot XI, original
configuration, in flight.*

Contents

Foreword

I had always hoped that some individual or organisation would one day issue a series of monographs on the classic machines of early aviation history. It is unfortunate that most of the celebrated aircraft of World War I and after have been thus honored, but not the great aeroplanes upon which all the rest have been based. So I am particularly happy that Dr. Crouch has started off with one of the truly "greats," the Blériot XI.

This famous aircraft had to be modified from its beginnings with an REP engine, primitive propeller, and its notable baby keel area supported on slender struts over the centre section. The 24-hp Anzani, the sophisticated Chauvière propeller, and a respectable amount of fuselage keel area completed the *desiderata* for a famous aeroplane. But it is often forgotten that a great aeroplane should be both an aeroplane which accomplished great deeds and one which had a great influence on history. The Blériot XI fulfills both criteria ideally. She accomplished the first cross-Channel flight on July 25, 1909, which sent an appalling shaft of fear through the military and naval commanders of the day, who foresaw the ultimate eclipse of sea power; and she influenced the design of countless monoplanes up to, and into, World War I.

Luckily, the original aircraft today survives to remind the French nation of early aeronautical glories. It is fitting that an American authority has courteously reached across the Atlantic to enshrine her in this admirable volume.

CHARLES HARVARD GIBBS-SMITH
First Lindbergh Professor of
Aerospace History at the
National Air and Space Museum
and a Fellow in the Science
Museum, London.

Acknowledgments

The author accumulated far more than the usual number of scholarly debts during the course of work on this volume.

C. H. Gibbs-Smith, Charles Dollfus, and Michel Lhospice have laid the groundwork for all inquiry into the history of Blériot Aéronautique. Donald Lopez, Chairman of the Aeronautics Department of the National Air and Space Museum, offered the encouragement and working atmosphere that made the task of researching and writing the volume enormously easier than would otherwise have been the case.

Dr. Howard S. Wolko, Special Advisor for Technology in the Aeronautics Department, was virtually the intellectual coauthor of the technical portions of the book. He offered invaluable assistance, particularly in unraveling the mystery of Blériot wing collapse. Melvin B. Zisfein, NASM Deputy Director, offered advice and assistance on aerodynamics. Brig. Gen. Benjamin S. Kelsey, USAF (Ret.) drew the answers to the author's interminable and frequently naive questions from his own considerable experience as an engineer and pilot of the aircraft of the wood, wire, and fabric era.

Claudia Oakes, a colleague in the NASM Aeronautics Department, puzzled through particularly difficult French translations with the author and remained ever willing to listen to the latest bit of Blériot trivia with a patient smile. Karl Schneide contributed his enthusiasm for and experience with the two Blériot aircraft maintained by Cole Palen at Old Rhinebeck Aerodrome.

Robert Mikesh, who superintended the restoration of the Domenjoz Blériot, offered his comments on NASM restoration policy. John Cusak, Joseph Fichera, Karl Heinzel, and Robert Pagett not only applied their skills to the restoration of a priceless piece of aeronautical history, but described their techniques to the author.

Catherine Scott, Mimi Scharf, Phil Edwards, Pete Suthard, and the other members of the NASM Library staff served as peerless guides through the treasures of their own collection, gathered material from other libraries, and assisted in collecting photographs. Leonard O. Opdycke, editor of *World War I Aeroplanes*, deserves special thanks for assistance with the photos. The Rozendaal drawings of the Blériot XI were also provided by Mr. Opdycke. G. Delaleau and the staff of the Musée de l'Air also provided photo assistance and opened their files to the author. Dale Hrabak, James Vineyard, and Lonny Rabjohns copied faded photographs with their usual skill.

C. H. Gibbs-Smith, Howard Wolko, Donald Lopez, Walt Boyne, Claudia Oakes, Dr. Richard Hallion, E. D. "Hud" Weeks, Leonard Opdycke, Phillip Jarrett, Harvey Lippincott, and John Bagley offered useful and detailed comments on the text. Dorothy Cochrane typed the manuscript with remarkable precision, patience, and interest (and laughed in all the right places).

Finally, my thanks go to my wife, Nancy, and our children, Bruce, Abigail, and Nathan, who were willing to accept a pater familias more at home in the Paris of 1909 than in modern Washington, D.C.

Blériot XI

Introduction

Lt. E. Bague of the French Army reaches the island of Gorgona, near Corsica, during an unsuccessful attempt to cross the Mediterranean in a Blériot XI on March 5, 1911.

Since the early seventeenth century the French have referred to the narrow sleeve of water that separates their nation from England as La Manche. It was Mare Britannicum to the Roman legions that crossed it in 55 B.C. The English seamen who turned back the Armada in 1588 knew it as Oceanus Brittanicus, the "British See." The term *English Channel* came into use during the eighteenth century, perhaps as a result of its identification as a "canal" in early Dutch navigation atlases, or perhaps to distinguish it from the "German Sea," as the North Sea was sometimes known.

By whatever name, these treacherous tide-swept waters have helped shape the course of European history. The Channel was England's moat, a barrier so successful that not since 1066 has an invading army crossed it to establish a beachhead on English soil.

Precisely because the Channel was so formidable an obstacle, it was also an irresistible challenge to the adventurous souls who dreamed of tunneling beneath it, swimming across it, or flying over it. The possibility of a cross-Channel tunnel was conceived as early as 1802. Work was begun, and then halted late in the nineteenth century, not to be revived until 1957. In 1872, J. B. Johnson became the first man to attempt to swim the Channel. Capt. Matthew Webb created something of a sensation in England when he completed the first successful swim on August 25, 1875.

But the prospect of a flight across the Channel, completely escaping the turbulent waters that had held Napoleon at bay, was particularly appealing. In January 1785, Jean-Pierre Francois Blanchard and Dr. John Jeffries caused a wave of excitement when they made the first crossing in a balloon. By September 1908 at least fifty-three other men and women had followed suit, completing a grand total of thirty-six cross-Channel balloon voyages. In 1903 Samuel Franklin Cody, an expatriate American aeronautical expert with the British Army, even made the crossing in a canoe towed by a large kite that he had constructed.

But this early aerial activity had little impact on English confidence in the inviolability of their Channel moat. The balloon was little more than a plaything for wealthy sportsmen, a helpless captive of the winds, that could scarcely be regarded as a serious military threat.

The next aerial crossing, the thirty-seventh, would be a quite different matter, for it would not be made by a balloon, but by an airplane. Louis Blériot's epic cross-Channel flight of July 25, 1909, constitutes one of those rare benchmarks in the history of aviation. Longer flights had been made, both in terms of the distance covered (23.6 miles) and the time in the air (36½ minutes), but nothing captured the public imagination in England and on the Continent in quite the same fashion. Not until Charles Lindbergh soloed the Atlantic in May 1927 would another aviator strike the same responsive chord.

The Blériot XI that made the flight was a primitive craft, to be sure. With its open box-frame fuselage, exposed pilot position, and a three-cyl-

Roland Garros thrills a crowd of French spectators.

inder engine that developed only twenty-five horsepower, it was scarcely more of a threat than the balloons that had preceded it.

Yet for all of its apparent fragility, the classic cross-Channel Blériot monoplane was the most significant and influential and the longest-lived aircraft design of the era. Between July 1909 and August 1914, over 800 Blériot aircraft were produced, most of them Type XI derivatives. A Blériot sales brochure prepared in 1910 listed the names of 104 individuals who had purchased Type XI airplanes and proudly called attention to the fact that the embryonic air forces of the French, British, Italian, Austrian, and Russian governments were flying Blériot aircraft. Flying clubs as far afield as Saigon and Sebastopol were operating Blériots as well. In 1911 Claude Grahame-White and Harry Harper listed 158 Type XIs in the hands of private pilots and noted that it was the most popular monoplane ever produced.[1]

A rare aerial photograph of a Paris/Rome-type Blériot XI.

These machines dominated the air-racing and exhibition circuit in the years prior to World War I. Major speed, distance, and altitude prizes offered by the sponsors of flying meets in Reims, Rouen, Bournemouth, New York, Brussels, Caen, Lanark, Nantes, and Bordeaux in 1909, 1910, and 1911, were captured by Blériot pilots.

Alfred Leblanc and Emile Aubrun, both flying Type XIs, took first and second place in the first great long-distance air race, the Circuit de l'Est, sponsored by *Le Matin* in 1910. Ensign Jean Conneau won three of the great 1911 distance races, Paris to Rome, the Circuit of Europe, and the Circuit of Britain, in his Blériot. The high-water mark came in July 1910 when the Type XI held the world's records for altitude, distance, speed, and duration.

As a result of the enormous success of the French-built originals, foreign manufacturers paid a premium for the right to produce Blériots under license. In addition, countless other less exact, and usually much less airworthy, copies were built by smaller firms and individuals in Europe and America.

As the first genuinely successful tractor monoplane, the Blériot was the direct forebear of a series of ever-more-sophisticated racing and military aircraft produced by firms like Morane-Saulnier, Blackburn, Fokker, and Deperdussin prior to 1914.

Clearly the Blériot XI was the most important of the classic aircraft that emerged during the magic summer of 1909. It remains one of the best-known and universally recognizable airplanes of all time. Yet at the time of its public unveiling on Christmas Eve 1908, there was little to suggest that the underpowered little monoplane was destined for immortality. In fact, as the authoritative publication *Flight* remarked, "it was not unusual to find doubts expressed as to its capacity for flight at all."[2]

Christmas Eve 1908: A huge airship dominates the First Paris Aeronautical Salon.

I

"Un Aviateur Militant"

Armand Fallières, eighth president of the Third French Republic, opened the second half of the Annual Paris Automobile Salon in the Grand Palais at 1:00 P.M. on December 24, 1908. The first section of the show, which featured the racing and pleasure cars from which the entire exhibition drew its name, had closed the previous week. The "deuxième série" now unveiled, included trucks, industrial vehicles, motorboats, and what was proudly billed as the First Paris Aeronautical Salon. Any doubts that the planners of the Salon might have had as to which section would prove most popular with the public were quickly laid to rest. As one observer noted: "The flying machines are, without the shadow of a doubt, the main attraction for everyone."[1]

The opening day crowds were so large that President Fallières and his distinguished entourage had difficulty viewing the exhibits. The sponsors of the show were forced to hire extra guards to control the 120,000 visitors who poured into the Grand Palais each day, crowding around the booths maintained by individual exhibitors to gawk at the machines on display and ply the salesmen with questions. Multilingual guides led throngs of foreign tourists through the exhibits, while the Compagnie Générale de Navigation Aérienne, which marketed Wright machines in France, went so far as to retain a representative who could extol the virtues of their product in Esperanto, the international tongue.

The exhibition, which was open for only six days, despite repeated requests that it be extended, was an enormous financial success. Admission fees as high as $1.00 during peak hours produced total revenues twice as high as those for the automobile portion of the salon, which had remained open three times as long.

The First Paris Aeronautical Salon deserved the attention it received. It was a historic event, the first large-scale display of aircraft, both practical and impractical, ever held. For months, news of flying machines and the intrepid pilots who flew them had dominated headlines. Now, at last, visitors could get a close look at these wonderful constructions of wood, wire, and fabric.

The Astra airship *Ville de Paris*, 184 feet long, 33 feet in diameter, and sporting twin Hotchkiss machineguns, towered over the entrance area. But the airship and the half-dozen free balloons surrounding it were overshadowed by the airplanes scattered through the remainder of the richly decorated hall.

A full-scale model of the Wright airplane was the star of the show. Wilbur Wright created a sensation when he paid a personal visit to the Salon on December 29. During the course of the exhibition the French syndicate selling Wright aircraft claimed to have received deposits on twenty-eight machines.

Virtually all of the emerging French aircraft builders were present as well. Monoplanes, biplanes, and helicopters produced by Léon Levavasseur, Alberto Santos-Dumont, Robert Esnault-Pelterie, Gabriel and Charles

An Antoinette VII (on floor) and a Voisin (above) in the 1908 Salon.

Voisin, Louis Breguet, Clément-Bayard, Alfred de Pischoff, and Raoul Vendôme also were exhibited.

The Blériot firm was well represented, with three aircraft on display. The new Blériot X attracted most of the attention. Clearly based on the Wright pattern, the three-place biplane was the largest machine Blériot had ever constructed. The Blériot IX, a large monoplane powered by a 50-hp, 16-cylinder Antoinette engine, featured wings covered with vellum paper, a standard feature of all Blériot models since the Type V of 1907.

The third machine in the Blériot pavilion, dwarfed by its larger and more impressive companions, drew much less notice. Although officially designated the Blériot XI, its narrow wingspan of only seven meters quickly led to its informal identification as the "short-span monoplane." A 30 to 35-hp, 7-cylinder, semiradial REP engine set behind a square bedstead support frame on the nose of the small craft drove a four-bladed metal propeller.

The Blériot X.

The Blériot IX during unsuccessful
flight trials at Issy-les-Moulineaux.

One of the first photos of the Blériot XI, taken at the Paris Salon, 1908. The tail of the Blériot X extends to the right with the Blériot IX in the background.

Louis Blériot (1872–1936).

Unlike many of the aircraft on display, it was apparent that much thought had gone into the control system of the Blériot XI. A stick located between the pilot's knees was moved from side to side to "warp" or twist the wings for lateral control. A fore-and-aft motion of the stick operated the pivoting elevator tips of the horizontal stabilizer, while a foot bar controlled the movement of the small rudder at the rear of the machine. A long, flat, fabric-covered tear-drop ran between the two upper-wing support pylons to help prevent side-slips.

The craft rested on a relatively heavy undercarriage that featured castoring front wheels mounted on spring shock-absorbers attached to the bedstead frame at the front. A large tail-wheel was also included.

While the aeronautical cognoscenti visiting the Paris Salon viewed the "short-span" monoplane with interest and curiosity, there was much doubt as to the performance of an airplane that sported only twelve square meters of wing area. Flight was a new venture, after all, and many of the aircraft on display (including the Blériot IX and X) would never leave the ground. Clearly, judgment on the Blériot XI would have to be withheld until the craft had flown.

At the time of the First Paris Aeronautical Salon, Louis Blériot was thirty-six years old. Physically he was a striking man, sturdily built, with a dark face and heavy features. His sweeping moustache, clear deep-set brown eyes, and high cheekbones led more than one commentator to remark on his resemblance to the popular image of an ancient Gallic chieftain. Frederick Collin, his mechanic, called attention to his "patron's" prominent nose and jocularly wondered if Blériot's "birdlike" profile might not be evidence of predestination.

Blériot was born in the northern industrial town of Cambrai on July 1, 1872. He was well educated, passing though the normal progression of schools from the Institute Notre-Dame in Cambrai, to the lycée at Amiens, to the Collège Sante-Barbe in Paris. At age twenty-three, immediately after graduating with a degree in engineering from the Ecole Centrale des Arts et Manufactures, he founded La Société des Phares Blériot, a firm specializing in the production of acetylene headlamps and accessories for automobiles.

In 1901, he married Alice Vedène, whom he had met while performing his military service as a lieutenant of artillery. As the first of their six children began to arrive, Blériot seemed to be settling into the position of a prosperous small industrialist.[2]

But the young businessman had already been infected by the flying bug. Blériot had apparently become interested in the possibility of heavier-than-air flight while at the Ecole Centrale, but had kept his enthusiasm to himself "for fear of being taken for a fool." Now, earning an average of 60,000 francs a year from the sale of headlamps, he could afford to indulge in his first aeronautical experiments.

Blériot made his initial forays into aeronautical design between 1900 and 1902, when he constructed as many as three unsuccessful model ornithopters powered by lightweight carbon-dioxide-gas engines. His choice of the antiquated flapping wing at so late a date is an indication of the low state of European aeronautics at the turn of the century. The recent well-publicized, well-funded efforts of Hiram Stevens Maxim in England and Clément Ader in France had failed to produce a successful flying machine. Otto Lilienthal and Percy Pilcher, Europe's leading glider pilots, were both dead by 1899, the victims of glider crashes. In the wake of these events, enthusiasts like Blériot, who continued to recognize the potential of heavier-than-air flight, were a distinct minority.

For Europeans it was the age of the airship. In France attention focused on Alberto Santos-Dumont and his small gasbags, while all Germany beamed with pride at the achievements of Count von Zeppelin. English aeronautical thinking had scarcely moved beyond the free balloon.

All of this was to change after 1902, as incomplete and frequently garbled

Blériot, his wife, Alice, and five of their six children.

accounts of the gliding experiments of Wilbur and Orville Wright began to reach Europe. The news was particularly disquieting to the French, who now seemed in danger of forfeiting a rich tradition of aeronautical leadership that dated back to the Montgolfiers.

Primarily as a result of the activity of Capt. Ferdinand Ferber, commander of the 17th Alpine Battalion at Nice, and Ernest Archdeacon, a wealthy Paris lawyer, a small band of heavier-than-air flying-machine enthusiasts had begun to coalesce by 1903 in the Aéro-Club de France. Determined to recapture aeronautical hegemony for their nation, these men built and flew gliders based on their understanding of the work of the Wright brothers, Octave Chanute, and the Australian Lawrence Hargrave, inventor of the box kite. Initial progress was slow as the French attempted to retrace the experimental steps of the Wrights without fully understanding the central importance of the control problem, but a foundation of enthusiasm and experience was laid that would enable French aviators to surpass all of their rivals in the years after 1909.

Louis Blériot was a leading member of this emerging circle of aeronautical enthusiasts. In 1905 the authoritative journal *L'Aérophile* aptly referred to him as "un des aviateurs militants de l'Aéro-Club de France." And militant he was, eagerly participating in group discussions and anxious to begin active experimentation on his own.[3]

On June 8, 1905, Blériot joined the small crowd of spectators lining the banks of the Seine between the Billancourt and Sèvres bridges to watch Gabriel Voisin test fly a new experimental float glider. Voisin, a twenty-five-year-old ex-student of architecture, was serving as chief engineer (at a salary of 190 francs a month) for the Syndicat d'Aviation recently organized by Archdeacon. The float glider, which incorporated box-kite wing-and-tail surfaces with vertical partitions and a forward elevator inspired by Wright practice, was the third machine Voisin had constructed for Archdeacon. It was a large craft, with 538 square feet of lifting surface, mounted on two long floats.

By 3:00 P.M. the glider had been towed to midstream and attached to the speedboat *La Rapière*. Voisin, secure on his "saddle," ordered the tow boat into motion. "I had the controls ready," he recalled fifty years later. "I waited for a time and then I applied [the] elevator. My glider instantly left the water. In a few seconds I was as high as the tops of the poplars along the quay." Approaching the Sèvres bridge, *La Rapière* slowed and the glider sank back to the water after a flight of some 2,000 feet.[4]

Blériot, impressed by the performance of the Voisin-Archdeacon glider, ordered a copy for himself on the spot. Early the next morning he appeared at Voisin's shop with suggestions as to how the basic Archdeacon configuration might be altered in his own glider. He asked that the wings be given a camber of one tenth (that is, a distance from the chord to the apex of the arch of one-tenth the length of the chord). Voisin feared that such a deep camber would lead to lateral instability, but he followed the judgment of the machine's purchaser.

By July 19, 1905, Voisin was ready to give the Archdeacon glider a second trial and make preliminary flight tests with the new Blériot glider as well. The Blériot II (Blériot I being the 1900–1902 ornithopter) closely resembled the Archdeacon original, but there were some significant modifications. The lower wing was shorter than the upper, so that the side curtains closing the outer wing bay were angled out. The camber of one-tenth had been applied to the biplane tail surfaces as well as the wings. Finally, the craft was a good deal lighter than the Archdeacon glider.

The Blériot II, constructed by Gabriel Voisin, on the Seine, July 10, 1905.

L'Antoinette, a motorboat designed by Léon Levavasseur, was chosen on this occasion to tow the two gliders. Because of its direct fuel-injection system, the Antoinette engine could not be controlled as well as the Panhard that had powered *La Rapière*. As a result, on the test of the Archdeacon glider the boat took off down river "like a mad thing," forcing a halt not long after Voisin had left the water. Safely ashore, Voisin decided to proceed gingerly with the test of the Blériot II.

Once again the boat surged ahead, drawing the glider into the sky. But, as Voisin had feared, the craft was unstable. It rocked violently from side to side and finally entered the water with its left wing after a flight of only one hundred feet. Voisin was barely able to extricate himself from the sinking wreckage with the assistance of a boatman.[5]

Far from discouraged by the near disaster, Blériot decided that the time had come to enter aeronautics in a serious fashion. Three days after the abortive trial on the Seine he offered to enter into a partnership with Voisin, who accepted immediately, eager to leave the ranks of hired mechanics and launch his own business. Their firm, the Blériot-Voisin Company, was the first in the world founded solely to produce airplanes.

But Voisin was soon to discover that there could be little doubt as to who was the senior partner. Blériot would allow him much less freedom than had Archdeacon in the design of their joint machines. Shortly after they had launched this venture, Blériot announced that their first project would

The Blériot III, Lake Enghien.

be the construction of a powered machine with tandem cylindrical wings of equal dimension. Voisin was stunned, though he had little choice but to conduct model studies of the proposed configuration. Gradually, he was able to convince Blériot that the wings should be elliptical rather than circular, but he could extract no further concessions.

According to Voisin, Blériot was also responsible for the powerplant decisions. He chose a 24-hp, 8-cylinder Antoinette engine that drove two tractor propellers, each 2 meters in diameter, at 600 rpm. The complex transmission, including twin flexible drive shafts, was produced by a contractor and weighed 243 pounds on delivery.[6]

The elliptical tandem wings of the craft provided sixty square meters of lifting surface and were constructed of hollow ash covered with varnished French silk. Twin elevators were housed in the front wing and a rudder in the rear ellipse. The machine was mounted on three pairs of lightweight wooden floats fitted with rubber flotation bags.

Work on the Blériot III continued from the fall of 1905 through the early spring of 1906. Voisin was particularly disturbed. "What a lot of time and money we lost" he recalled many years later.[7]

Finally, in late May 1906, the glider was transported to Lake Enghien, near Paris, for flight trials that Voisin characterized as "disastrous." The heavy glider refused to rise from the surface of the lake.

Blériot was now willing to relent, suggesting that the front ellipse be replaced with a standard box-kite biplane wing with twin side curtains. A Wright-derived forward biplane elevator provided pitch control. Two narrow ailerons were mounted at the midpoint of the trailing-edge struts on the forward wing.

Blériot's use of the aileron was, as C. H. Gibbs-Smith has noted, interesting but scarcely original. He paid very close attention to the work of his colleagues and undoubtedly borrowed the notion from Robert Esnault-Pelterie, who had incorporated ailerons on an unsuccessful glider in 1905.

Two 24-hp Antoinette engines were directly linked to twin contrarotating propellers. The empty weight of the machine was 430 kg.

Like its predecessor, the new Blériot IV was mounted on floats and shipped to Lake Enghien for testing. During trials held on October 12 and 18, 1906, the craft raced up and down the lake at speeds of up to 30 kph, but would not fly.[8]

By mutual consent Blériot and Voisin returned to the shop at Billancourt and modified the craft for takeoff from land. On November 12, their simple alterations completed, they trucked the Blériot IV, now sporting wheels, to the open grounds made available to aviators at Bagatelle, an eighteenth-century estate north of Longchamp on the west side of Paris's famous Bois de Boulogne.

The machine ran ineffectually up and down the field. On the final run, with a mechanic at the controls, the Blériot IV hit a stone and bounced across a shallow ditch, seriously damaging the front surfaces and breaking the propellers. While there was some thought of rebuilding the machine with a single 50-hp Antoinette replacing the twin 23-hp engines, it was clear that there was little hope for the Blériot IV.[9]

While November 12 was an unhappy day for Blériot, it was regarded as "miraculous" by most of the aeronautical enthusiasts at Bagatelle. Following the morning trials of the Blériot IV, Alberto Santos-Dumont, a diminutive Brazilian expatriate who had become the hero of all Paris as a result of his flights in a series of small dirigible balloons, wheeled out his latest creation, 14-bis, for an attempt to capture an Aéro-Club de France prize of 1,500 francs for the first flight of over one hundred meters.

Like the 1905 Voisin float gliders, 14-bis was a crude combination of Wright (pusher biplane with canard elevator) and Hargrave (box-kite wing with side curtain) technology. Powered by a 50-hp Antoinette, it made its first free flight on September 13. On October 23, Santos-Dumont had covered some fifty meters, winning the Coupe d'Aviation Ernest Archdeacon for the first flight of over twenty-five meters. Now, on November 12, with the dejected Blériot and Voisin in the audience, he nursed 14-bis through a series of hops, the best of which, 200 meters (722 feet) in 21⅓ seconds, won the Aéro-Club prize.

It was a turning point for Blériot. Having seen a man fly, he was even more determined to get into the air himself. Both Blériot and Voisin were convinced that their association had proved fruitless and should be dissolved. Voisin would enter into a new and much more successful partnership with his brother Charles, while Blériot struck out on his own, establishing a new shop on the Avenue Victor Hugo.

It was a new beginning for Blériot in another sense as well. He had found it far too frustrating to stand aside and watch Voisin or a mechanic pilot his machines. From now on Blériot would do his own flying.

By January 1908, "Le Patron," as he was known to his workmen, had begun to gather a staff. Louis Peyret, an experienced modelmaker who had served as a mechanic for Blériot-Voisin, was elevated to foreman of the

The Blériot V.

new establishment. Young Louis Paragot, "Petit Louis," was only fourteen when he signed on as the world's first aeronautical apprentice. Raymond Saulnier performed general engineering duties, while Ferdinand Collin, fresh from military service, joined the staff as a mechanic. Alfred Bertrand was hired to supervise construction. M. Granseigne was a joiner and mechanic, while Julien Mamet, M. Pelletier, and a few others assigned to general duties were to join the staff by mid-1909. In addition, a number of Blériot's friends, notably Alfred Leblanc, an experienced balloonist, could be counted on to lend a hand around the shop when required.

This closely knit team of talented individuals deserves far more credit for Blériot's success than they have received. During the first decade of powered flight, public attention focused on the pilot, the intrepid fellow who risked his life to taste the freedom of the skies. The craftsmen who created the machines, the men who had solved the myriad technical problems that had to be overcome before these primitive craft could be coaxed aloft, often remained in the background, their names and achievements unknown to the public.

This attitude was reflected in the most knowledgeable circles. *Flight*, the most authoritative English-language publication in the field, remarked as late as 1909 that the designers and constructors of airplanes "deserve a degree of credit for their work, which is far higher than the uninitiated are apt to accord them, although their names must of course always stand second to those intrepid pioneers who have actually practised the art of flight."[10]

As both chief pilot and owner of the firm, Blériot received all of the credit for the performance of the machines that bore his name. In fact, he was often responsible for only the most general design decisions.

Voisin regarded Blériot as "an engineer of genius," yet remarked that his old partner, and others like him, did not really design their own machines.

"These . . . pioneers relied entirely on the instinctive abilities of their assistants to build their machines—obviously under the general direction of themselves," he later contended.[11]

Aircraft design was an empirical, cut-and-try process during these pioneering years. The eye of the craftsman, the ear of the mechanic, were more important tools than the slide rule or drawing board. Ferdinand Collin recalled that "not once have I held in my hands the smallest drawing or sketch of Blériot's, neither I nor any of my coworkers. We were driven hard to invent and construct parts with nothing more than the objective indicated, and that only in general terms."[12]

When visiting the Blériot booth at the Grand Palais in 1908, Gustave Eiffel, the distinguished engineer who had designed the tower that had become a symbol of Paris, asked Collin to explain the "scientific formulae" used to guarantee the stability of the Blériot machines. The embarrassed mechanic was forced to reply that they "consisted simply of shifting the position of the pilot's position or altering the location of the wings to the front or rear to correct the balance if the airplane had a tendency either to dive or climb."[13]

Given the informal nature of Blériot's design process, the importance of the workmen and mechanics becomes apparent.

With a staff assembled, work began on their first project, the Blériot V. Clearly Blériot had been much impressed by the hops of 14-bis, for his new machine was obviously inspired by Santos-Dumont's configuration. Like 14-bis, it was a canard, with the engine, a 24-hp Antoinette, placed at the rear of the craft driving a pusher propeller. The cockpit was located in the triangular fuselage immediately in front of the engine. A single surface elevator and a rudder were placed on the nose.

Blériot departed from his earlier practice by using swept-back monoplane wings (covered with vellum paper) with a span of 7.8 meters and a chord of 2 meters at the root. The wings, swept back and up, the forward tail, and two light bicycle tires that served as an undercarriage, gave the machine a rakish, sinuous appearance.

The pilot was provided with several controls and even an instrument, a spirit level. A universally jointed hand lever on the pilot's right operated both elevator and rudder. A second control stick enabled him "to raise and twist either the right or left wings in order to turn." While contemporary accounts are unclear as to the exact function of this "raising and twisting" action, it sounds suspiciously as though Blériot was attempting to apply the Wright wing-warping system to the new machine.[14]

Blériot was now only one step away from developing the "modern" control system, in which a single stick controls motion in pitch and roll, while a foot bar operates the rudder. Between December 1906 and January 22, 1908, Robert Esnault-Pelterie, another French pioneer, applied for three patents covering a system very similar to that envisioned by Blériot. Blériot filed independently less than a month later, on February 9, 1907. While priority in the development of this psychologically natural control arrangement has always been clouded, Blériot deserves a very real share of the credit for the innovation.

Unfortunately, to test a flight control system one must have a machine capable of flight, and in that regard No. V was not much of an improvement over the earlier craft. Tested at Bagatelle on March 21 and 27 and April 2, 5, 7, 15, and 19, it suffered a series of discouraging crashes, punctuated by several short hops of from two to five meters.[15]

Blériot's next venture, No. VI, was a nearly complete departure from

*Louis Peyret in the cockpit of the
Blériot VI.*

The Blériot VI.

A heavily retouched photo of the Blériot VI in the air.

anything he had tried before. At this stage of his career, he was casting about rather wildly for a configuration with sufficient promise to warrant development.

The new machine was a tandem monoplane derived from the Langley Aerodrome of 1903. Louis Peyret, who had begun testing tandem-wing flying models as early as 1903, was primarily responsible for the design of this craft. It seems quite likely that the instability noted in the short hops that spring led Blériot to approve the construction of a machine with generous wing area and radical dihedral that promised much in terms of inherent stability.

When complete, the Blériot VI had 18 square meters of surface area. As in the case of the Blériot V, parchment paper was chosen as the covering material. Blériot believed that paper offered fewer problems in damp weather, decreased air resistance, and made for quicker repairs than fabric. The trusty 24-hp, 8-cylinder Antoinette turned a single tractor propeller.

The Libellule (Dragonfly), as the Blériot VI was dubbed, demonstrated that Blériot and his staff were still wrestling with the control problem. The standard rudder linked to a control stick was employed, but there was no rear elevator at all. To control pitch, Blériot relied on the manipulation of a spark lever, the use of wing-tip elevators, and weight shifting.

Unsuccessful flight tests were held at Bagatelle at 5:00 A.M. on July 7, 1907. Following these trials Blériot enlarged the wings and reduced propeller pitch.

He was back at Bagatelle on July 12 and made a flight of thirty meters. Three days later, on July 15, he covered eighty meters, flying "as high as a second story." In spite of damage to the propeller and undercarriage, Blériot was elated. He had covered less than the distance of a football field, but, for the first time, he had actually flown.

Realizing that pitch stability remained a serious problem with the craft, Blériot had ordered the seat moved forward eighty centimeters. A flight test on July 24, when the tail rose but the machine refused to lift, demonstrated that he had gone too far.

When the Blériot VI was brought out for its next trial on July 27, the seat was mounted on rollers "as in skiffs," so that the pilot could shift his weight fore or aft as required. The first flight of the day was a straight line hop of 120 meters. During the second flight Blériot made his first tentative turn in the air, negotiating a 200-meter semicircle that ended when he accidentally moved his seat forward too abruptly.

With the prospect of longer flights, Blériot shifted operations to Issy-

The Blériot VII, first of Blériot's tractor monoplanes.

les-Moulineax, a large military parade ground on the Left Bank, just south of the Boulevard Victoire. As early as March 1905, Voisin and Archdeacon had used the field to test gliders towed by automobiles. The influential Vuia monoplane had been tested there in August 1906. By 1909, "Issy," as it was affectionately known, would become the most famous flying field in the world.

On August 1, Blériot flew 100 meters in 6.5 seconds at Issy. He obtained an even better flight, 265 meters, on August 6. On this occasion he covered the first 122 meters at an altitude of 2–3 meters, then momentarily touched down before rising to 12 meters to fly on. The flight ended abruptly when Blériot moved too far forward once again.[16]

Following a short flight of 80 meters on August 10, Blériot substituted a 50-hp Antoinette for the 24-hp model originally used. Other modifications, including a reduction in the dihedral and tail surface, were made before flights resumed in September. Six more flights were made with No. VI that month, the best covering 184 meters. The end came on September 17 in one of those apparently disastrous crashes for which Blériot was to become famous. "The monoplane took off like an arrow," he reported. "Everything went well at first. Very quickly I was up to 25 meters. I was quite impressed with this height when all at once the motor stopped dead. The machine started down. Believing myself lost, I left my seat and threw myself back toward the tail. This maneuver almost worked; the craft leveled off, lost speed, and slowly crumpled to earth."[17] The Dragonfly was destroyed, but Blériot had escaped injury and had learned a valuable lesson that would be applied on a number of subsequent occasions. "A man who keeps his head in an aeroplane accident is not likely to come to much harm," he later commented. "What he must do is to think only of himself, and not of his machine; he must not try to save both. I always throw myself upon one of the wings of my machine when there is a mishap, and although this breaks the wing, it causes me to alight safely."[18]

The Libellule was a turning point in Blériot's fortunes. In spite of the manifold problems of the machine and its unhappy end, it had enabled Blériot to get into the air for the first time. Capt. Ferdinand Ferber, who had been the first Frenchman to recognize the value of gliding experiments, commented on the importance of this factor in Blériot's case. "Let us remark. . . ," he wrote in 1907, "how fruitful is the method of personal trial which we have always advised in preference to any calculation. This year, with his fourth [*sic*] apparatus, Blériot has not met with any damage to his aeroplane. [Well, not much anyway.] He made the trials himself and they quickly led to results, because each trial gave him an exact idea of what was to be corrected. That is the condition of success."[19]

Blériot's next machine, No. VII, which debuted at Issy on October 5, 1907, was a tractor monoplane. The rectangular fuselage tapered to a point at both ends. The wings, spanning 11 meters, had a surface area of 25 square meters. A 50-hp Antoinette drove the four-bladed metal propeller.

The wings were given a dihedral angle to provide automatic lateral stability. A rudder and elevators at the tips of the horizontal stabilizer were the only moveable control surfaces. The total weight was 425 kg.

Early trials indicated that the undercarriage was too light. After strengthening and the completion of flight tests with the Blériot VI, the new machine was back at Issy on November 7, 15, 22, and 23, making no appreciable flights and suffering a series of minor crashes.

Finally, on November 20, Blériot achieved a flight of 150 meters. By December he had stretched his best distance to 200 meters. Three days later, he was covering as much as 500 meters through the air. On December 18, at about 3:05 in the afternoon, Blériot was landing when the left wheel hit a hole and collapsed, flipping the airplane completely on its back. When the dust cleared the machine was destroyed, but Blériot, "the valiant sportsman," dug himself out of the wreckage having suffered only minor contusions. He was already gaining something of a reputation for being accident-prone—and lucky.[20]

Blériot and his staff had to be content watching other men fly while they prepared their next project, No. VIII, during the early spring of 1908. Blériot must have had some fear that he would be left behind during these months. On January 13, Henri Farman flying a Voisin biplane captured the

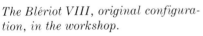

The Blériot VIII, original configuration, in the workshop.

The Blériot VIII, June or July 1908, at Issy. Note the triangular ailerons, engine mount, and foreshortened nose.

Grand Prix d'Aviation Deutsch-Archdeacon for the first official mile flown in Europe. By April 1908, while the Blériot team was still completing their latest model, pilots like Léon Delagrange were keeping Voisin machines in the air for over five minutes at a time.

The Blériot VIII made its first appearance in April, when it was photographed in the shop for *L'Aérophile*. The fuselage, rectangular in the center section, was drawn to a point in front and a triangle at the rear. The 50-hp Antoinette was completely enclosed as in the Blériot VII. A square rudder, elevators, and small triangular ailerons were employed.

For the first time the classic Blériot cockpit controls—stick and rudder bar—appeared. The "cloche," a bell-shaped piece of metal attached to the bottom of the control stick, operated control lines leading to the elevators and ailerons.

The wings were set at a dihedral angle and had a span of 11 meters, 80 centimeters. The total surface area was about 25 square meters.

In general terms the new craft looked very much like its predecessor, but the details of construction indicated that the long winter spent in the shop had not been wasted. The undercarriage, which had given Blériot so many difficulties, was completely redesigned. The basic bedstead framework, which doubled as an engine support and the major structural element for the undercarriage, was in place. The presence of upper and lower pylon braces for the flying and landing wires was evidence of a new concern for the loads and stresses to be encountered in flight.

After moving rapidly through a series of wildly divergent designs inspired by the Wrights, Hargrave, Santos-Dumont, and Langley, Blériot at last developed a design that was uniquely his own. On his eighth try he had produced an airplane capable of genuine flight on which he could both gain experience in the air and test various control techniques and refinements.

In fact, the airplane was modified so frequently between April and July 1908 that it is difficult to date the changes precisely. The first alterations were made before the aircraft was flown. As noted, photos taken in the shop in April show the Blériot VIII with the fuselage completely enclosed from tip to tail. The fuselage longerons were drawn to a point some distance in front of the engine, requiring a long propeller shaft. No control surfaces were visible on the wings.

The Blériot VIII-bis. Note wingtip "flap" ailerons.

The final version, Blériot VIII-ter, in flight. Note pivoting ailerons.

Photos taken at Issy during the first flight trials in June and July show a bewildering variety of changes. The rear half of the fuselage and the engine are exposed. The forward ends of the longerons have been cut off at the front of the engine, and a much shorter propeller shaft is in place. Two types of ailerons are shown. Presumably, a set of very small triangular wing-tip ailerons were fitted first, perhaps even at the time of construction. By July 20, these ineffective surfaces had been replaced by large, square, flap ailerons. By August 1, the ailerons had been entirely removed and center-pivoted wing-tip ailerons added. A large horizontal stabilizer had also been placed on top of the fuselage framing forward of the elevator, which was positioned on the underside of the frame.

Blériot referred to the final version as the VIII-ter (third) and to that fitted with flap ailerons as the VIII-bis (second).

From the outset of flight tests on June 17, the new craft seemed to have marked the end to Blériot's legendary bad luck. By July 3, with ailerons

An accident with the Blériot VIII-bis.

Damage to the tail after a flight on October 22, 1908.

installed, he was flying his first complete circles and remaining in the air for over eight minutes at a time. He extended his flights during the summer, teaching himself to fly while at the same time trying the various improvements on his machine.

By October 9, with the new wing-tip ailerons mounted, Blériot was reaching altitudes of up to fifteen meters. On October 21 he was awarded a 2,500-franc altitude prize offered by a group of sportsmen.[21]

Ten days later Blériot made one of the first round-trip flights between two towns. The day before, on October 30, 1908, Henri Farman had flown his modified Voisin from his flying field at Camp Chalons to Reims, a distance of twenty-seven kilometers.

On October 31, Blériot made what his contemporaries regarded as "a

Blériot; his wife, Alice; Alfred Leblanc (right); and André Fournier (commissioner of the Aéro-Club de France) plan the Toury-Artenay flight.

still more sensational and perfect journey," covering twenty-eight kilometers from Toury to Artenay and back. Following a takeoff at 2:50 that afternoon, he flew over Château Gaillard at an altitude of twelve meters. Eleven minutes after his start he was forced down with ignition trouble. With the assistance of the team following him in automobiles, he repaired the magneto and was on his way again after an hour and a half on the ground. On the return journey he was forced down a second time at Villiers Farm, near Santilly, and then proceeded back to Toury.

The French press was ecstatic. As one contemporary remarked: "Louis Blériot thus demonstrated that the French aeroplanes mounted on wheels are complete apparatuses, truly self-starting, practical, and capable of resuming their flight when it is interrupted; he showed the services that aerolocomotion could render us, illustrated that aviation from that time henceforth could enter into everyday practice."[22]

With the Toury-Artenay flight, the first portion of Louis Blériot's aeronautical career was drawing to a close. The goals that he had set in 1905 had been achieved. Blériot had joined the very exclusive ranks of the men who could fly. Yet it had cost him dearly to join this exalted company. The 760,000 francs (roughly $150,000) spent during the course of his experiments had exhausted his resources, including his wife's dowry. Now, in the wake of the triumphant flights of October, Blériot's financial situation was deteriorating rapidly. A heavy investment in his next two machines was completely lost when the Blériot IX proved too heavy to fly, while the Blériot X was so obviously inadequate that it was apparently never tested.

Facing financial disaster, Blériot nevertheless forged ahead with two new airplanes, the "last chance monoplanes" as he dubbed them. "I had to keep going," he would recall. "I couldn't do anything else. I had to keep going because, like a gambler, I had to recover my losses . . . I had to fly."[23]

The Blériot XI, original configuration, in flight.

II

"C'est un Triomphe!"

ith the close of the Salon at the Grand Palais on December 30, 1908, Blériot entered a period of frantic activity, desperate to recoup his financial losses with prize money and aircraft sales. Everything depended on the success of his two new monoplanes, the Blériot XI and XII.

As with all Blériot machines, the story of the design of these new craft, particularly the XI, is something of a tangled web. Charles Dollfus, the eminent French aeronautical historian, who knew Blériot and his leading workmen well, points out that the young engineers who came to work for the firm brought their own plans for aircraft with them and were free to develop these under Blériot's general guidance. Louis Peyret, for example, had been primarily responsible for the tandem-wing Libellule, a configuration that had fascinated him for several years.

In the same manner, Dollfus argues that the Blériot XI was, for the most part, the work of Raymond Saulnier. He notes that Blériot took little initial interest in the development of the most famous of all his machines. During late 1908, when the Blériot XI was under construction, Blériot's attention was focused on his successful flights with No. VIII and the work on Nos. IX and X.[1]

Still, it seems unlikely that Blériot played no role in the development of his new craft. The Blériot XI incorporated many of the elements that had evolved through the earlier models, including the three-wheel undercarriage with the front wheels mounted on a shock-absorbing bedstead; the pylon supports for the wings; the rectangular "trellis" fuselage, uncovered at the rear; the small rudder and pivoting elevators; and the now-standard Blériot control arrangement.

Moreover, the use of wing warping for lateral control was almost certainly one of Blériot's personal contributions. He had been present at the Hunaudières race-course near Le Mans on August 8, 1908, when Wilbur Wright made his first European flights. The *New York Herald* quoted his reaction to Wilbur's performance. "I consider that for us in France, and everywhere, a new era in mechanical flight has commenced. I am not sufficiently calm after the event to thoroughly express my opinion. My view can be best conveyed in the words, 'It is marvelous!' "[2]

Ross Browne, an American who had attached himself to the Blériot camp and had accompanied the group to Le Mans, recalled Blériot's enthusiasm many years later. "Blériot was all excited; he looked over the machine . . . he tested the wings, and Mr. Wright showed how the warping was done. . . . how it worked."

Following the flights of August 8, Browne remembered that Blériot had remarked: "I'm going to use a warped wing. To hell with the aileron." He was, Browne recalled, "just like a young boy." We need look no further for the origin of the wing-warping system of the Blériot XI.[3]

Thus, while the Blériot XI was undoubtedly Saulnier's project, it was a natural extension of earlier Blériot machines and one to which Louis Blériot had made substantial contributions.

The REP engine and the fabric teardrop to prevent side-slips are apparent in this photo of the original Blériot XI in flight.

The Blériot XI, February 1909.

Preparing for takeoff at Issy.

The Blériot XII at Reims, August 1909.

The Blériot XI made its first flight, a short hop of 200 meters, at Issy on January 23, 1909. While the little craft seemed stable enough, Blériot realized that the wing area of only twelve square meters was simply too small. When it made its next appearance on February 18, it sported fourteen square meters of wing surface. Two flights of 700 meters each were made that day in a brisk wind.

Encouraged, Blériot announced that he was transferring operations to the larger flying field at Buc, near Versailles, where Robert Esnault-Pelterie, who had constructed the engine mounted in the Blériot XI, conducted his own flight tests.

At Buc between March 9 and 15, Blériot made flights of up to 2,500 meters with turns. By early April he was making daily flights of up to two and three minutes, gradually getting the feel of the machine.[4]

On May 21 he brought the second of his new monoplanes out for its first trial. Much larger and heavier than the XI, the Blériot XII was a high-wing aileron-equipped monoplane powered by a 35-hp ENV engine. Throughout the remainder of the spring and early summer of 1909, Blériot was to fly both machines regularly.[5]

When the Blériot XI made its next appearance on May 27 it had been altered yet again. The large 7-cylinder REP, which Blériot considered too heavy for the machine, had been replaced by a much smaller and lighter 3-cylinder, 25-hp Anzani powerplant.

Alessandro Anzani, who had designed and built the engine, was an Italian bicycle racer with a feisty temper and, as Ferdinand Collin recalled, the manners and vocabulary of a trooper. Sometime earlier, Anzani had begun the manufacture of motorcycle engines and was persuaded by two employees, a German named Hoffman and an Italian mechanic, Francesco Santarini,

Blériot XII. Note tail group.

to add a third cylinder. Santarini also suggested punching holes at the base of the cylinders to assist in cooling the new engine.

Compared with the craftmanship of the beautiful fuel-injected Antoinettes with which Blériot was familiar, the Anzani certainly didn't look very impressive. Its three iron cylinders were rough castings, with seventeen holes drilled in the base to exhaust spent gases. In later years Collin recalled the enormous amount of play in the Anzanis and commented that they "rattled and spat oil out of the holes and the end of every stroke, covering the pilot with a greasy film, so that one had to be heroic and long suffering to keep flying with these miserable engines."[6]

Yet the Anzani had one great virtue: it ran continously for up to an hour at a time. The details of Blériot's original agreement with Anzani are unclear, but apparently the engine that first powered the aircraft in May and was still in place for the Channel flight was only on loan. Blériot hoped to produce the engine' to accompany his airframes and had signed an agreement for exclusive rights with Anzani, but he had been unable to pay the Italian the agreed-upon price.

In July 1909 Alice Blériot was able to free her husband of his financial obligation to Anzani. While visiting friends she saved the young son of a wealthy Haitian planter from a serious fall. The grateful father, who was also an aeronautical enthusiast, advanced the money with which Blériot cleared his debt for the engine and licensing agreement, opening the way for the Channel flight.

But in May 1909, all of this lay two months in the future. Blériot's immediate problem was to demonstrate the worth of his aircraft, and this he did in no uncertain terms. By late June both the XI and XII were performing superbly, and Blériot was making flights of up to thirty-seven minutes. Clearly he had overcome the technical problems and required only a spectacular flight to popularize his machines and open a market for his airplanes.

Early in July he shipped the XII to Douai for a meet at the Brayelle aerodrome, while the XI shuttled between Issy and Port Aviation, a new flying field at Juvisy, some twelve miles south of Paris. Blériot himself traveled back and forth, taking part in concurrent meets at both sites.

Blériot's luck seemed to be changing. Late in June he was named to share the 100,000-franc Osiris Prize with Gabriel Voisin. This was a major award

The Anzani engine and Chauvière propeller, spring 1909.

The cross-Channel Blériot XI from Victor Loughead's Vehicles of the Air *(Chicago, 1911).*

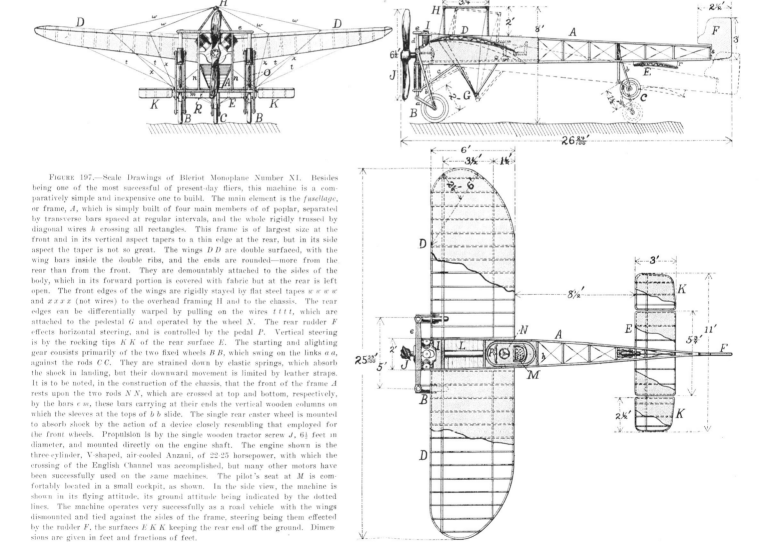

FIGURE 197.—Scale Drawings of Bleriot Monoplane Number XI. Besides being one of the most successful of present-day fliers, this machine is a comparatively simple and inexpensive one to build. The main element is the *fusellage*, or frame, *A*, which is simply built of four main members of of poplar, separated by transverse bars spaced at regular intervals, and the whole rigidly trussed by diagonal wires *h* crossing all rectangles. This frame is of largest size at the front and in its vertical aspect tapers to a thin edge at the rear, but in its side aspect the taper is not so great. The wings *D D* are double surfaced, with the wing bars inside the double ribs, and the ends are rounded—more from the rear than from the front. They are demountably attached to the sides of the body, which in its forward portion is covered with fabric but at the rear is left open. The front edges of the wings are rigidly stayed by flat steel tapes *w w w w* and *x x x x* (not wires) to the overhead framing H and to the chassis. The rear edges can be differentially warped by pulling on the wires *t t t t*, which are attached to the pedestal *G* and operated by the wheel *N*. The rear rudder *F* effects horizontal steering, and is controlled by the pedal *P*. Vertical steering is by the rocking tips *K K* of the rear surface *E*. The starting and alighting gear consists primarily of the two fixed wheels *B B*, which swing on the links *a a*, against the rods *C C*. They are strained down by elastic springs, which absorb the shock in landing, but their downward movement is limited by leather straps. It is to be noted, in the construction of the chassis, that the front of the frame *A* rests upon the two rods *N N*, which are crossed at top and bottom, respectively, by the bars *e m*, these bars carrying at their ends the vertical wooden columns on which the sleeves at the tops of *b b* slide. The single rear caster wheel is mounted to absorb shock by the action of a device closely resembling that employed for the front wheels. Propulsion is by the single wooden tractor screw *J*, 6½ feet in diameter, and mounted directly on the engine shaft. The engine shown is the three-cylinder, V-shaped, air-cooled Anzani, of 22–25 horsepower, with which the crossing of the English Channel was accomplished, but many other motors have been successfully used on the same machines. The pilot's seat at *M* is comfortably located in a small cockpit, as shown. In the side view, the machine is shown in its flying attitude, its ground attitude being indicated by the dotted lines. The machine operates very successfully as a road vehicle with the wings dismounted and tied against the sides of the frame, steering being them effected by the rudder *F*, the surfaces *E K K* keeping the rear end off the ground. Dimensions are given in feet and fractions of feet.

Hubert Latham: "Mr. President, I am a Man of the World."

offered once every three years by the prestigious Institut de France for "the most important discovery or work in the department of Science, Literature, and Art, or in anything that may be of conspicuous public interest."[7]

Blériot's luck in the air was improving as well. He won a number of major prizes and, at Douai on July 3, flying the Blériot XII, he remained in the air for forty-eight minutes, covering forty-two kilometers. Even this occasion was marred, however, when the lining of the exhaust pipe blew out, causing third-degree burns to Blériot's left foot.

Eleven days later he was back in the air, embarked on yet another distance flight. This time he was out to win a government-subsidized prize of 14,000 francs offered by the Aéro-Club de France for the first straight-line flight of over forty kilometers to be completed in six hours.

He had chosen to fly the XI from Etampes to Orléans, partially retracing the route of his most important flight of 1908. On the evening of July 12, the airplane was towed to the takeoff field some 1.5 kilometers south of Etampes. At 3:30 the next morning Blériot and his party, which included his wife, Alfred Leblanc, and André Fournier, drove to the site. Shortly thereafter the Blériot XI was in the air after a 50-yard run, and three automobiles carrying the rest of the party were racing along back-country roads in an attempt to keep the airplane in view.

Near Bormainville, having passed over a trainload of excitedly waving passengers, Blériot set his craft down, to demonstrate, he later remarked, that it was possible to stop and restart en route. Ten minutes later he was off again, passing Toury, Château Gaillard, and Dabron, over which he had flown the previous year in No. VIII. Finally, with an early morning wind rising from the west, he made a difficult landing at Croix-Briquet-Cheville, thirteen kilometers from Orléans. The time was 5:44 A.M. He had covered forty-two kilometers in just under forty-five minutes.

The prize money from the Orléans flight and the Osiris Prize were of some help in easing Blériot's financial problems. Blériot himself realized 9,000 francs from the flight (5,000 as pilot and 4,000 as constructor), while Anzani was awarded 3,000 francs, and Lucien Chauvière, who designed and built the Intégrale propeller, received 2,000 francs. But the real importance of the flight was in the new self-confidence it gave Blériot. He now felt ready to accept a greater challenge. He would fly the Channel.

More than any other individual, Alfred Harmsworth, Lord Northcliffe, was responsible for the enormous excitement that surrounded the Channel flight. An aeronautical enthusiast and patriot, Harmsworth had become alarmed at the rapid progress being made in the United States and France while Britain remained inert. In an effort to rouse British interest in flight, he established a number of major aeronautical prizes for distance flights, races, and model competitions in the years prior to 1914.

Harmsworth's most important challenge was issued by his newspaper, the *London Daily Mail*, on October 5, 1908. He offered £500 ($1,250, later raised to $2,500) to "the first person who shall succeed in flying across the Channel from a point on English soil to a point on French soil, or vice versa."[8] The flight had to be made on a heavier-than-air machine that, between sunrise and sunset, could cover the distance without intermediate landings in the water.

At the time the prize was announced, Harmsworth was fully aware of the fact that Wilbur and Orville Wright were the only men capable of flying the Channel. To encourage them to make the flight he privately offered Wilbur an additional $7,500, three times the original prize money. But the Wrights refused this challenge as they had all others, preferring to concen-

Latham (left) and Léon Levavasseur.

*Latham and Antoinette Gastambide,
for whom the Antoinette aircraft and
engines were named.*

trate all their energies on demonstrations that were directly linked to ne-
gotiations for the sale of their machines.

Not until June 1909 did Hubert Latham announce his intention of trying
for the prize. In an era dominated by eccentric and individualistic pilots,
Latham was the most colorful of the lot. A native Parisian, he was born on
January 10, 1883, to an extraordinarily wealthy family which maintained
English citizenship in spite of the fact that they had lived in France for
three generations. Latham had spent fifteen months at Oxford's Balliol
College, long enough to pass his law examinations before beginning his two
years of required service as a private in the French Army, stationed at St.
Cloud.

Following his military service, Latham found himself unable to settle into
a business career. He became a world traveler and adventurer, leading
safaris to Africa, racing automobiles and motorboats, and ballooning across
the Channel with his cousin, Jacques Faure. Several years later, when
President Fallières asked Latham what his profession was, he could only
reply: "Monsieur le Président, je suis un homme du monde."

Latham had been drawn into aeronautics though the efforts of his neigh-
bor, Jules Gastambide, an electrical manufacturer whose chief engineer,
Léon Levavasseur, had pioneered the development of lightweight aeronautical engines. Born in Cherbourg in 1863, Levavasseur was the son of a
petty officer in the French Navy. He had come to Paris at age seventeen
to study at the Beaux Arts but had become so intrigued by electricity that
he abandoned art for engineering in 1880. He had developed a variety of
arc lamps, generators, and alternators before turning to aeronautics.

As early as 1903 Levavasseur had produced an unsuccessful flying ma-

Takeoff, July 19, 1909.

chine, but his real achievement during the years before 1905 was the creation of the earliest Antoinette engines. Light and powerful, about 1 kg per horsepower in the 16-cylinder version, the Antoinettes were named after Gastambide's daughter. Initially they were used to power the famous *Antoinette* racing boat that took first prize in the Monaco races in 1904, 1905, and 1906 (and unsuccessfully attempted to tow the Blériot II into the air in 1905). The novice aviators of the period were quick to recognize the potential of the Antoinette as an aeronautical powerplant. Santos-Dumont, Ferber, Voisin, and Blériot had all relied on these engines during their early experiments. Blériot himself was named vice-president of the firm established to produce Antoinettes in 1906.

By 1908 Levavasseur had produced his first successful aircraft, the Antoinette II (sometimes Gastambide-Mengin II). This aileron-equipped craft, a modification of the Antoinette I that made its first appearance at Bagatelle in February 1908, was the first monoplane to fly a complete circle and the first to carry a passenger.

Levavasseur's next craft, Antoinette III, was constructed for Captain Ferber. The Antoinette IV, which followed in October 1908, was the first of the classic Levavasseur designs, a graceful monoplane with large wings, a narrow fuselage, and beautifully sculpted ailerons and empennage.

In February 1909 Levavasseur and Gastambide invited Latham to learn to fly the Antoinette IV. After a few brief hops with Eugène Welferinger, the only other pilot to have flown the machine, he was on his way. On June 5, 1909, no longer a novice, Latham kept the Antoinette IV in the air for 1 hour 7 minutes 37 seconds, establishing a new French duration record and a world record for monoplanes. The following day he flew 6 kilometers in 4 minutes 13 seconds to capture the Ambroise Goupy prize. On June 12 he demonstrated the Antoinette before government officials, covering a distance of 40 kilometers in 39 minutes.

Almost overnight Latham became one of the most popular pilots in Europe. With his natty attire topped off by an ever-present checked cap, cigarette holder poised at a rakish angle, he was the very image of the intrepid aviator.

During the months since the announcement of the *Daily Mail* prize, a number of aviators, including Santos-Dumont, Charles Stewart Rolls, and Count Serge de Bolotoff, had been mentioned as possible contenders for the award, but Latham was the first to arrive on the scene with a machine and begin preparations for the flight.

He would take off from the cliffs at Sangatte, a small village some six miles from Calais. His camp consisted of a tent pitched against the wall of the old tunnel works abandoned after Lord Grosvenor's abortive attempt to tunnel beneath the Channel in 1883. Latham himself lodged at the Grand Hotel in Calais, while Levavasseur, his son, and ten Antoinette mechanics stayed at the Hôtel de la Plage in Sangatte, closer to the airplane. And so they waited until the weather broke, amusing themselves with the out-of-tune piano at the Plage.

The French destroyer *Harpon* had been placed at Latham's disposal. In addition, the *Daily Mail* had rented the tug *Calesian*, equipped with a crane, to follow the Antoinette across. To add to the excitement the *Daily Mail* had hired the Marconi Co. to establish a wireless station near Latham's hangar, so that the aviator could have a constant source of information on the English weather, while the *Daily Mail* kept a close watch on activity in the Antoinette camp.

As rain and high winds grounded Latham at Sangatte, the Channel cross-

The Antoinette IV at Sangatte.

ing began at last to take on the aspect of a race. Count Charles de Lambert, an émigré Russian who had been Wilbur Wright's first pupil in France, announced his intention to compete, choosing for his two Wright airplanes a takeoff site near Wissant, south of Sangatte. But the fates were to conspire against de Lambert. One of his machines was damaged in a trial flight, and the other was not yet ready by the time of the Channel crossing.

Early on the morning of July 13, 1909, as Blériot was winging his way to Orléans, Latham decided to attempt his first flight since arriving at Sangatte. As there was no suitable spot for takeoff near the hangar, the Antoinette was tugged up a narrow hilly lane to an open field behind the cliffs at Blanc Nez. Latham climbed rapidly and circled the village of Sangatte in a strong wind, then steered back to the field. In landing he missed a clover field and descended into a patch of standing corn. While the pilot was uninjured, the right wheel and landing support-strut of the airplane were broken.

On July 15 conditions again seemed ripe for the Channel attempt, but when the airplane was wheeled out it was discovered that one set of storage

Latham, outfitted in a hat and coat borrowed from his rescuers, returns to Calais on July 19.

batteries had discharged overnight and the reserve set had been stolen. The seven hours required to recharge the batteries meant another day lost.

Four days later, on July 19, Latham was finally off. Only six miles out his engine stopped, with the English shore not yet in sight. He glided into the water just 500 yards from the *Harpon*. Although the engine pulled the nose under, the aircraft remained afloat, and Latham, clad in a blue turtleneck pullover and a tam-o'-shanter, discovered that his pockets were still dry and so proceeded to light a cigarette while waiting for a rescue.

Latham was accorded a hero's reception in Calais. Ludovic Bréton, the chief tunnel engineer, welcomed him back in the name of the citizens of the town, remarking: "You are my rival, you wish to go over, while our aim was to pass under La Manche. But you have done magnificently, mon cher." A "buxom, red-cheeked fisher girl," the queen of the port, gave him a kiss as the crowd cheered "Vive Latham.!"[9]

Both Latham and Levavasseur took this initial failure well. Latham had the distributor component that had failed made into a stickpin while Levavasseur was reported as being even prouder of his Antoinette. "We have a machine which can go on land, in the water, and in the air," remarked "Père Antoinette." "It runs, it flies, it swims. C'est un triomphe!" Wilbur Wright took a more jaundiced view of the proceedings, commenting that he "was surprised the Antoinette got as far as it did."[10]

Latham's failure decided the issue for Blériot. The Antoinette IV was so severely damaged that it would have to be completely rebuilt. Another

Antoinette, this time a wing-warping model VII, would be sent immediately to Sangatte, but at least a week might be required to prepare it for flight. Count de Lambert still seemed in no hurry to make the flight. Fresh from his triumphant flights of early July, Blériot saw the opening he had waited for.

Immediately after the conclusion of the Orléans flight on July 13, Blériot ordered his machine broken down. Thirty-nine minutes later, the wings neatly folded to the side and the tail wheel resting in a truck bed, the Blériot XI was on its way to the shop at Neuilly, near Paris. Over the next week Blériot made occasional flights, fine-tuning his machine and awaiting the outcome of Latham's attempt. During one of these trials on July 18 his left foot was burned a second time. Now hobbling about on crutches, Blériot discovered that he could still operate the rudder pedals and, following Latham's failure, he determined to rush into the breach.

The Blériot XI arrived in Calais on a railroad flatcar covered with striped canvas early on the morning of July 21. All morning curious crowds milled around the station to catch a glimpse of the machine. The citizens of Calais who were accustomed to the graceful, sweeping lines of Latham's Antoinette were perhaps a bit disappointed. As one correspondent noted: "It is a monoplane less than half the size of Mr. Latham's, but very similar to it in design. Seen close, it is not beautiful, being dirty and weatherbeaten, but it looks very businesslike."[11]

Blériot himself arrived on the scene that afternoon, having undergone minor surgery on his foot only the day before. After registering at the Terminus Hotel in Calais he immediately began a search for a suitable takeoff site. He rejected the grounds of the Calais Casino, finally settling on a farm at Les Baraques, between Sangatte and Calais. Unlike Latham, who once again planned to take off from the cliff of Blanc Nez, Blériot had chosen a low pasture that led directly over the beach.

The engine for Latham's new Antoinette VII arrived in Calais with the Blériot XI. The new airframe arrived at noon on Thursday, July 22. Levavasseur immediately set ten mechanics to work assembling the craft. By Friday morning Latham felt secure in offering the required forty-eight-hour notice of his intention to make a second try. That evening Comte de Lambert officially entered his own name in the list. Now, at last, a real Channel race was developing.[12]

It was a race in which the weather, and not an official timer, would determine the start. Excitement built as high winds grounded all of the aviators from July 21 to 24. Correspondents reported Blériot, who was ready to go, as praying for a break in the winds, while Latham and Levavasseur, racing to assemble an airplane that had never been flown, hoped for continued bad weather.

While newspapers touted the "Exciting Contest for the *Daily Mail* Prize" and speculated on "The Excitement at Calais" and the "Channel Rivals," Blériot and Latham were arranging the final details. There was the matter of escort vessels for example. In the event that both men left at the same time, the captain of the torpedo boat assigned as a rescue craft declared that he would be obliged to follow Latham, the original contestant. As a result, a second vessel, the destroyer *Escopette*, was assigned to Blériot by the government.

Reporters were also gathering in force. Naturally Lord Northcliffe intended to capitalize on his investment by headlining the story. Harry Harper, the *Daily Mail* correspondent in Calais, assumed he would use the Marconi station to scoop the world.

Enterprising French journalists were just as determined not to be out-done. On July 21 Charles Fontaine, a reporter for *Le Matin*, had been ordered to accompany Blériot to Calais and render all possible assistance, while at the same time feeding constant bulletins to the paper. Fontaine, his friend and fellow reporter Robert Guérin, and M. Maes, a *Le Matin* photographer, were quick to make themselves useful. They accompanied Blériot, Alfred Leblanc, Collin, and Mamet on the search for a takeoff site and assisted in assembling the airplane.

Fontaine then left for Dover on the Channel boat *Pas-de-Calais*. His mission was to search for a suitable landing site and to arrange for a photographer to be on hand at the end of the flight. Fontaine had every intention of being the first reporter to talk to Blériot after the great event.

Arriving in England, the reporter traveled to the famous Shakespeare Cliffs near Dover, which, rising some 100 meters above the Channel, he knew would be too high for Blériot to cross. While motoring around the countryside he found a small flat area, North Foreland Meadow, to the right of Dover Castle, quite near the spot where Blanchard and Jeffries had lifted off for the first balloon flight across the Channel. Here was a natural entry without the forbidding cliffs that made up so much of the Channel coast.

Fontaine immediately sent a picture postcard of the area to Blériot with an X marking the low entrance to North Foreland Meadow, less than 100 feet above the water. In an accompanying letter he promised to be at the site on the morning of the flight, waving a large tricolor to guide Blériot in.

Back in Calais, there was little to do but wait for an improvement in the weather. Both the Blériot XI and Antoinette VII were now ready for flight. Collin had rigged a large rubber flotation bag in the open fuselage of the Blériot to help support a pilot who could not hope to swim with his injured foot. In addition, he had sighted the Dover light one evening with a small compass, marked the course with a copper wire laid within the pilot's view, and taped the compass to the airplane. In later years Blériot would be unable to recall even having seen the compass during the flight.[13]

The tension of waiting was beginning to tell on Blériot. After an engine test on July 24 he turned to Mamet and Collin and asked: "What do you think, Mamet, will I succeed?" When Mamet remained silent, he repeated the question to Collin. Perhaps taken aback, the mechanic suggested that he shouldn't go if we was not confident. Blériot snapped back: "I'm not asking about myself. I'm asking what you think about the machine!"[14]

With some hope of improved weather, both Blériot and Latham arranged to be awakened early on the morning of Sunday, July 25. Alfred Leblanc would be responsible for rousing the Blériot camp if a flight seemed possible. Levavasseur would do the same for Latham.

Leblanc was up at 2:10 the next morning. The skies were clear. The wind had died at last. He ordered the car out and went to wake Blériot and his wife at 2:30.

Blériot's mood had not improved since the day before. His foot hurt. He was nervous and refused to eat breakfast. In his own words: "I would have been happy if they had told me that the wind was blowing and no attempt was possible."[15]

Nevertheless, the small group emerged from the hotel a half hour later. Dressed in tweed and protected from the chill by blue coveralls and a khaki jacket, Blériot was beginning to revive. Alice Blériot was escorted aboard the *Escopette*, and then the party proceeded to Les Baraques, where Anzani, Collin, and Mamet were preparing the Blériot XI for flight. Anzani had awakened all of them with typical exuberance an hour or so earlier by firing blanks from a pistol.

By 4:00 A.M. the engine had been thoroughly warmed up and Blériot climbed into the cockpit for a trial flight. Just before takeoff a small dog ran into the spinning propeller. Not the most propitious of omens.

After a fifteen-minute circle of Calais, Blériot was back on the ground, waiting for sunrise, when the *Daily Mail* rules would permit a takeoff for the Channel flight. While Collin and Mamet topped off the seventeen-liter gasoline tank, added castor oil to the engine, checked the flotation bag, and gave the airplane a final check, Leblanc trained a telescope on the Latham camp but could discern no activity.

Levavasseur, who had suffered a restless night, had overslept. As a result Latham was not up and about until well after Blériot's departure, when recurring high winds and the ineptitude of Levavasseur's ten mechanics prevented him from chasing his rival across the Channel.

Sunrise came at 4:35. Leblanc signaled the start and the little Anzani, operating at 1,200 rpm, strained to carry the Blériot XI up and over the telegraph lines at the edge of the small cliff. Once safely aloft, Blériot throttled back to save the engine. Traveling at roughly 43 mph at an altitude of 250 feet, he quickly left the *Escopette* behind. "The moment is supreme, yet I surprise myself by feeling no exultation. Below me is the sea, the surface disturbed by the wind, which is now freshening. The motion of the waves beneath me is not pleasant. I drive on."

Ten minutes out, Blériot was surprised to discover himself completely alone, out of sight of both shores and the pursuing destroyer. "For ten minutes I am lost. It is a strange position, to be alone, unguided, without compass, in the air over the middle of the Channel. I touch nothing. My hands and feet rest lightly on the levers. I let the aeroplane take its own course. I care not where it goes."

Twenty minutes after takeoff he caught his first sight of the English coast. "I was nearly safe. I steered toward the white cliffs. But the wind and fog caught me. I fought with my hands, with my eyes. I kept steering toward the cliffs, but I couldn't see Dover. Where the devil was I?"

Blériot headed south along the cliffs, following three Channel boats that he presumed were steaming toward the harbor. But the wind grew worse than ever. "Suddenly, at the edge of an opening in the cliffs, I saw a man energetically waving a tricolor . . . screaming bravo! bravo!"

As he had promised, Fontaine was marking the entrance to North Foreland Meadow. But there remained the problem of landing.

"Once more I turn my aeroplane, and describing a half circle, I enter the opening and find myself over dry land. Avoiding the red buildings on my right, I attempt a landing, but the wind catches me and whirls me round two or three times. At once I stop my motor, and instantly my machine falls straight upon the land from a height of sixty-five feet. In two or three seconds I am safe upon your shore."[16]

It was over. Louis Blériot had flown into history. He had covered the roughly 23 miles between Les Baraques to North Foreland Meadow in 36½ minutes, and in so doing had established himself as an aeronautical leader of the first rank.

Blériot was greeted by a few soldiers, a policeman, and Fontaine and his photographer. Fontaine ran up to help the pilot extricate himself from the badly damaged machine. Blériot's first question concerned Latham. When informed that his rival had not taken off, Fontaine noted that Blériot's face cleared noticeably. Latham cabled his congratulations and remarked that

The captain of the Escopette *(left), Alice Blériot, Louis Blériot, and Alfred Leblanc. Afternoon, July 25, 1909.*

Louis and Alice Blériot surrounded by well-wishers.

he would soon follow, but his disappointment was unmistakable. One reporter recalled seeing him not long after news of Blériot's success was received at Sangatte. His head was resting on the wing of the Antoinette as he brushed tears from his eyes.

Latham would attempt to follow Blériot two days later, on July 27. Once again he was forced to land in the water, this time suffering facial injuries when his head struck the aircraft. He remained one of the most popular pilots in Europe and America for several years, a familiar sight at meets with his jaunty cap and cigarette. He died on June 7, 1912, not in a crash, as one might expect, but gored to death by a wounded buffalo while on a hunting expedition near the Chari River in the French Sudan.

Moments after the landing, the crowd in North Foreland Meadow began to swell. First to arrive was the *Daily Mail* correspondent, driving a car with the Comte de Lapeyrouse, who had arrived from Calais the evening before, and a second bobby who had received word of the aviator's arrival. For the newcomers, it was an emotional scene. "And there was the Man. In the blue overalls of the French mechanic, the trousers torn, wearing a motor-cap of the kind that covers the ears and leaves the face looking like a mask, with one foot in a brown boot and the other in a slipper, M. Blériot stood beside his machine. His face was shiny with perspiration, but he seemed as calm as if he had just got out of a hansom cab. He was smiling. There was happiness, triumph, in his keen, dark eyes."[17]

Gradually, Dover awoke to what had happened. The small crowd grew as a dozen policemen and some soldiers, farmers, and children from the neighborhood gathered around the damaged airplane. "Everybody, I think, was excited to the point of incoherence," the *Daily Mail* correspondent recalled. "The Frenchmen were almost sobbing. M. Blériot was the calmest person there."[18]

These first few minutes were a forecast of what was to come. Recognizing this as a historic moment, early arrivals insisted upon autographing the craft. City officials, with Blériot's approval, quickly put up a tent over the airplane and charged sixpence a head admission, the money being donated to local hospitals and the police pension fund.

The London department store magnate, Gordon Selfridge, donated £200 to the London Hospital for the right to exhibit the craft for three days. Beginning on Monday morning, the day after the flight, huge crowds gathered at Selfridge's to inspect the now-famous Blériot XI. The display proved so popular that Blériot and the *Daily Mail* agreed to a one-day extension of the loan. On the evening of Thursday, July 29, Selfridge was forced to remain open until midnight to accommodate last-minute viewers.

Soon after the landing, the *Daily Mail* reporter and Fontaine hustled Blériot into a car and drove to the Lord Warden Hotel, with Fontaine's large tricolor held aloft from the auto. Refreshed, Blériot returned to the dock area to greet his wife and the cheering crew of the *Escopette*. The couple would return to Calais that day and be back in Dover once again on the night boat.

Everywhere they went during the next few weeks they were feted. Blériot, now a national hero in both France and England, was created a Chevalier of the Legion of Honor. In a very short time it became clear that Blériot and his monoplane had not made just another long flight. By crossing the narrow sleeve of water that had halted the troops of Napoleon, man and machine had become immortal.

The enormous reaction to Blériot's success is, perhaps, more than a little puzzling. Certainly longer flights had been made before without creating

Raising the Blériot XI for an exhibition at the offices of Le Matin *after the flight.*

the sense of excitement that Blériot elicited. Nor would any of the other great long-distance aerial voyages of the prewar era, including Roland Garros's epic flight over the Mediterranean in 1913, be cast in the same heroic terms in the eyes of the general public. Not even the great transatlantic flights of 1919 would have the impact of this one short early-morning hop across the Channel.

Not until 1927, when Charles Lindbergh soloed the Atlantic, would an aeronautical hero of Blériot's stature emerge. It should not be too surprising to note that the news photos of Lindbergh which many Frenchmen found most appealing were those which showed the lean, tousled young American being embraced by an aged and balding Louis Blériot, tears streaming down his cheeks.

In part, the impact of Blériot's flight was the result of the peculiar circumstances that had surrounded the Channel crossing. The dashing Latham had given it a gallant try and failed, but had maintained his unshakable aplomb. Then the crossing had been transformed into a race, with Blériot as a dark-horse entry. The tension had mounted while the aviators waited for a break in the weather; finally, a quick victory for one contestant while the other was left in tears. Journalists and the reading public found these elements irresistible.

Blériot's courage and perseverance were also extraordinarily appealing. It had, after all, been a difficult flight, as Latham's two failures had demonstrated. Charles Dollfus has underscored this point with his observation that "no pilot of today, no matter how great, could repeat this exploit in such an aircraft, with such an engine."

But Blériot's flight was far more than appealing journalism or an example of personal courage. Had that been the case, the event would have been forgotten in a few weeks or months. But by linking two nations across the most historically significant geographic barrier in Europe, Blériot's flight symbolized the coming of age of the airplane.

Perhaps the editor of *The Aero* came closest to capturing the essential meaning of the event when he observed that Blériot had "brought home to the minds of the people of all nations the possibilities of the flying machine in the future." To those with foresight and vision, the social and geopolitical implications of the airplane were suddenly brought into clear focus.[19]

For the English it was a rude awakening. The words of Admiral Lord St. Vincent negating the possibility of a French invasion in 1801 had long been the backbone of British defense policy. "I do not say they cannot come, My Lords, I only say they cannot come by sea."[20] But Blériot had come by air, and, as Sir Alan Cobham would later note, July 25, 1909, "marked the end of our insular safety, and the beginning of a time when Britain must seek another form of defense beside its ships."[21] The fact that "there was no more sea," as the *Pall Mall Gazette* observed, "betokened a revolution in human affairs."[22]

The flight had achieved Lord Northcliffe's goal. As he had insisted to the readers of the *Daily Mail*, there could be little doubt that "England is no longer an island."[23] Moreover, while French pride in their nation's aeronautical achievements was fully restored by the flight, England's backward position was also highlighted. H. G. Wells spoke to this point in Lord Northcliffe's *Daily Mail:*

What does it mean for us: One meaning, I think, stands out plainly enough, unpalatable enough to our national pride. This thing from first to last was made abroad . . . Gliding began abroad when our young men of courage were braving the dangers of the cricket ball. The motor car and its engine were worked

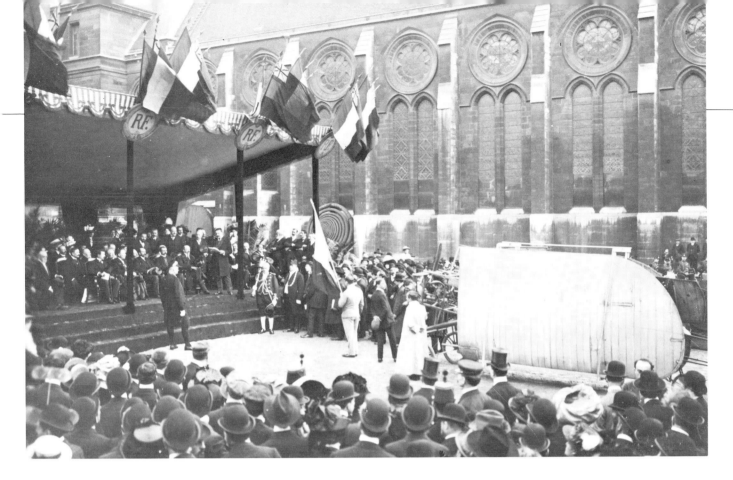

Presenting the Blériot XI to the Mu-sée des Arts et Métiers, Paris.

out over there . . . Over there where the prosperous classes have some regard for education . . . where people discuss all sorts of things fearlessly and have a respect for science . . . It means that the world cannot wait for the English. It is not the first warning we have had.

It has been raining warnings on us—never was a slacking, dull people so liberally served with warnings of what is in store for them . . . In the men of means and leisure there was neither enterprise enough, imagination enough, nor skill enough, to lead in this matter . . . Either we are a people essentially and incurably inferior, or there is something wrong in our training, something benumbing in our atmosphere and circumstances. That is the first and greatest intimation in M. Blériot's feat.[24]

Never again could the airplane be viewed simply as a mechanical marvel. In the wake of Blériot's flight preeminence in aeronautics was to become a matter of national honor—and survival.

The original cross-Channel Blériot XI is now on exhibit at the Musée des Techniques CNAM in Paris.

The classic Type XI, with an Anzani.

III

"Type Onze"

The months following the Channel flight were a time of transition for Blériot. Since 1905 he had struggled to develop a practical airplane and to learn to fly it. In the years after 1909 he capitalized on his fame to become the world's most successful aircraft manufacturer.

The lure of prize money and the opportunity to provide further demonstrations of the performance of his machine kept Blériot flying in the immediate aftermath of the Channel crossing. In August he was at Reims to participate in the world's first great week-long aviation meet. That fall he fulfilled engagements in several European capitals.

He delighted huge crowds in Budapest in mid-October, where he was introduced to the Archduke Joseph and his wife. The end of October found him in Vienna flying before a crowd of 300,000 spectators that included the Emperor Franz Joseph and the French ambassador. Early in November he was nearly mobbed by an angry crowd when engine problems prevented a flight in Bucharest.

Blériot's career as an exhibition pilot came to an abrupt end in Istanbul that December. He had gone aloft in spite of a strong wind, perhaps remembering the wrath of the disappointed Rumanians. Flying at an altitude of only sixty feet he was blown toward Tataola Hill, a nearby residential area. Unable to climb, he struck a house and fell twenty-five feet to the ground. He was able to extricate himself from the wreckage of this, his thirty-second crash, but complained of severe internal pain. He was taken immediately to the French Hospital in Istanbul, where he was diagnosed as having suffered several fractured ribs and minor damage to his spleen and liver. When complications set in he was transferred to a larger hospital in Vienna but was sufficiently recovered to return home to Paris by Christmas Eve.

It was Blériot's most serious accident, and his last. Contrary to reports that he ceased flying entirely after December 1909, he did continue giving lessons, testing factory aircraft, and taking his family aloft for an occasional joyride, at least until 1912, when he apparently grounded himself for good. But never again would he compete in a race or undertake a strenuous exhibition schedule. After 1909 the enormous energy that he had channeled into flying would be redirected into business.[1]

As early as September 1909 Blériot had informed a reporter for the *Echo de Paris* that he was considering giving up flying to concentrate on business. Prior to the Channel flight he had received fifteen orders for copies of the Blériot XI. By September, 101 airplanes were on order and Blériot was seriously worried about his ability to meet delivery dates.

Factory space was a central problem. The original facility at Neuilly was scarcely more than a shed. As the demand for airplanes rose sharply in the spring of 1910, Blériot purchased an additional building known as the "old Bowling Palace" in Neuilly, where he concentrated on production of the No. XII high-wing monoplane. Later that year the firm moved to entirely new quarters, a specially constructed factory on the Route de la Révolte in Levallois, near Paris. By late 1911, when the 500th Blériot airplane was

45

A Blériot XI in the air, summer 1909.

wheeled through the doors, more than 150 engineers and workmen were employed at the Levallois plant. While trustworthy aircraft production statistics are difficult to obtain for the pre-World War I era, it seems clear that Blériot was now preeminent in terms of the sheer number of machines built and sold.[2]

When the first two Blériot XIs had been sold to Alfred Leblanc and Léon Delagrange, an experienced Voisin pilot, the familiar Issy flying field had sufficed as a spot for flight instruction. By 1910, as aircraft production moved into the hundreds, new facilities for training Blériot pilots were required.

Three new flying fields were established. During the spring and summer months, activity centered at Etampes, a village on the plains of Beauce, some fifty minutes from Paris by rail. Each winter the school was transferred to Canbois, some six miles from Pau, in the south of France. A third school was established at Hendon, near London, for English-speaking students.

Novice aviators were to remember the "splendid aerodrome" at Canbois, near Pau, with particular fondness. The field was a 3½-mile circular track extending north of the Gane River on the Pont-Long plain. The area had been completely cleared of vegetation and rolled until it was reasonably smooth, a rarity among flying fields of the era. The area was fairly isolated, and a small village, complete with restaurant and hotel, had sprung up to serve the needs of the student pilots. Like Etampes, Pau was subdivided into military and civilian schools.

Flight instruction at the three schools was free for those who had purchased Blériot machines. For other students, the cost of instruction averaged 800 francs ($150.00) plus the cost of any damage to school aircraft.

Studies were broken into six units, running from ground-school classes on engine and airframe maintenance through extended practice at taxiing, to straight-line hops, turns, cross-country flights, and, finally, preparation leading to the award of a Fédération Aéronautique Internationale pilot's brevet.

The Blériot schools were extraordinarily successful. By mid-March 1912, a total of 200 civilian aviators, 61 French military pilots, and 16 foreign military aviators had received flight instruction at Pau or Etampes. This number included 16 Americans, 17 Italians, 6 Belgians, 6 Swiss, 3 Chileans, 12 Russians, 3 Swedes, 2 Peruvians, 3 Turks, 2 Rumanians, 2 Austrians, 2 Spaniards, a Mexican, a Greek, and a Dutchman.

Ensign Jean Conneau, a French naval officer who trained at Pau, vividly recalled his first impressions of life at the Blériot school:

> The series of attempts, expectations, disappointments, extends over seven days, and seven dull days they are, during which uncertainty is added to physical exertion. From time to time an incident breaks the monotony. I enjoyed inspecting the different machines, looking at them hopping on rough ground, zigzagging about like will-o'-the-wisps, then stopping all at once. Over yonder there is a machine rolling in a fantastic manner. Here a pupil alights, his face is black, his clothes, saturated with oil, give him the appearance of a demon; he takes them off, but leaves behind him as he goes along an unbearable smell. I assure you, learning aviation lacks poetry. Sometimes I would closely inspect the aeroplanes which are put at our disposal. Each of the parts . . . [has] been more than once replaced. One has a brand-new right wing and a dirty, oil-stained left one. Another machine is all the colors of the rainbow; a third one is a dirty yellow color.[3]

Earle Ovington, an American student at Pau, kept a diary in which he recorded the circumstances of his first solo:

> The grease-covered mechanics wheeled out one of the patched-up machines kept especially for 'taxi-drivers' like myself, and I clambered into the cockpit. The cane-bottom seat was not ten inches wide and its back consisted of a strip of three-ply veneer, three inches wide and a quarter of an inch thick. To make it still lighter, it was bored full of holes. The French certainly do peel down their machines to make them light. I had been told to steer for a pylon at the other end of the field, and as my little monoplane bumped unevenly over the ground, I must have concentrated too much on the pylon and not enough on what I was doing. I pressed my feet so heavily on the rudder cross bar that the back of the seat gave way, and I slipped over onto the bottom of the fuselage, pulling the elevator control to me as I went. Not realizing in the least what had happened, I scrambled back into position as quickly as I could. Instead of being on the ground as I supposed, I was three hundred feet in the air and still rising . . . Between wiggling the rudder with my feet, working the wings to keep the horizon where it belonged, and pushing and pulling the elevator to stop the earth from jumping up and down, I had a busy sixty seconds.[4]

With the factory and flying school operating smoothly, Blériot was free to resume aeronautical experimentation. In all, between 1909 and 1914 he produced some forty-five distinct aircraft types, ranging from canards reminiscent of the Blériot V to flying buses capable of carrying eight passengers. But for the most part, these types were produced in very small numbers and must be regarded as strictly experimental machines. (See Appendix C for a list of Blériot aircraft types, 1901–14.)

Throughout the period the classic Blériot XI, the "Type Onze," remained the firm's most successful product. Frequently altered and updated, these aircraft accounted for the lion's share of the rough total of 800 aircraft produced by Blériot Aéronautique through 1914.

The Type XI aircraft constructed during these years were marketed by the firm in four basic categories: trainers, sport or touring models, military aircraft, and racing or exhibition machines. In order to understand the various modifications of the basic Blériot XI pattern produced during the period 1909–14, it is helpful to take a closer look at each of these categories.

TRAINING MACHINES

The first Blériots sold were close copies of the original cross-Channel airframe-and-engine combination. The firm continued to produce these craft at least through 1913, although they were no longer first-line machines after 1910, having been superseded by improved models.

These Anzani-powered originals were sold in large numbers as "Appareils d'Ecole"—school machines. They were so simple to construct that in 1911 the factory could promise an average delivery time of only five days after an order was placed. Compared to other aircraft of the period, they were also relatively inexpensive, averaging only 12,000 francs ($2,352) in 1911–12.

Special "school machines" were sometimes referred to as "Penguins," a reference to the difficulty student pilots often experienced in coaxing them into the air. Severely clipped-wing versions incapable of flight were pro-

An early model XI with a Gnôme mounted in place.

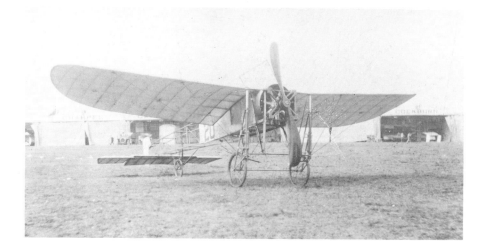

A Blériot "Penguin."

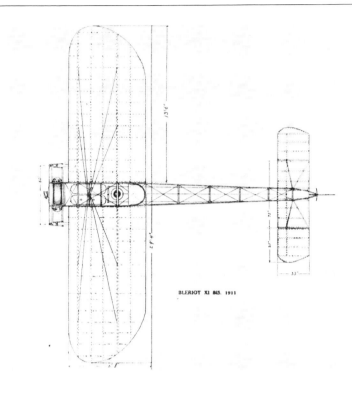

An uprated Blériot XI (1911).

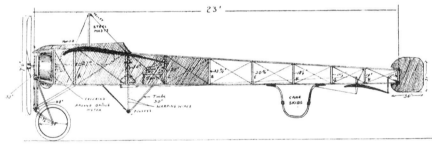

duced as military trainers just prior to and during World War I. (See Appendix C for details of a wartime Penguin.)

1911 BLÉRIOT XI
Appareil d'Ecole

Span:	8.9 m
Length:	7.65 m
Area:	15.0 m²
Weight:	220 kg (empty)
Speed:	65 kph
Gas Consumption:	12.5 liters per hour
Oil Consumption:	2.5 liters per hour
Engine:	25–30-hp Anzani

An advanced trainer powered by a 50-hp Gnôme rotary engine was produced after 1912. With a flight weight of 415 kgs, the uprated craft could reach 95 kph.

SPORT AND TOURING MODELS

The uprated Blériot XI, the firm's standard and most popular product, first appeared in 1910. The most important single change from the original cross-

Jacques Balsan, owner of the first factory original Gnôme-Blériot combination. Blériot in derby.

Channel Blériot was the substitution of a 50- or 70-hp Gnôme rotary engine for the Anzani. Léon Delagrange was the first to install a 50-hp Gnôme in a fuselage designed for an Anzani. The first Blériot XI with a factory Gnôme installation was purchased by J. Balsan in November 1910.

While the basic design remained the same, there were other modifications as well. After 1911 a long sloping metal hood covered the upper portion of the engine, tanks, and the forward section of the cockpit, providing some protection for the pilot and generally streamlining the machine.

The use of the Gnôme rotary necessitated mounting the engine behind the bedstead with cutouts on the fuselage sides to permit rotation. These were usually covered by convex metal panels extending either halfway or entirely down the fuselage side.

By 1911 the rear tail-wheel had been replaced by a Malacca or rattan tailskid that shortened the long landing-runs. The 1911 instrument panel also showed a marked improvement. An altimeter, clock, compass, angle-of-incidence indicator, engine-revolution counter, and map case were standard.

The tail surfaces were also gradually modified after 1910. The "high" rudder of the cross-Channel machine was replaced by a square rudder with rounded corners in the production versions. The pivoting elevators at the tips of the horizontal stabilizer gave way to a standard one-piece elevator by 1911–12. The dual tripod guyposts for the bracing wires had given way to single upper-and-lower pyramid supports.

The success of this uprated Blériot XI in leading meets and races during the years 1910–12 led to what were, for the period, enormous sales. (See Appendix A for a chronology of Blériot races and records.) One hundred and twenty of the craft had been sold by the end of 1910. A year later total sales had climbed to 300. The selling price for the 50-hp Gnôme version, fixed at 21,500 francs ($4,200) in 1910 rose to 24,000 francs ($4,700) in 1911. The price for the same aircraft fitted with the 70-hp Gnôme was 30,000 francs ($5,800). The promised delivery time of only eight days is testimony to the efficiency of Blériot production techniques.

1910–14 BLÉRIOT XI

Span:	8.9 m
Length:	7.8 m
Area:	14.0 m²
Weight:	240 kg (empty)
Speed:	90–100 kph
Tanks:	60 liters (gasoline)
	20 liters (castor oil)
Duration:	About 3 hrs. or 300 km (seven hrs. with extra tanks)
Engine:	50-hp Gnôme Rotary

The Blériot Type XXVIII Populaire, which first appeared in 1912, was a "bargain basement" version of the standard Type XI. While incorporating many of the advanced features of the uprated version, the Populaire was marketed with the old Anzani and was of generally lighter construction, with a reduced undercarriage and an empty weight of only 240 kg. The price was 11,800 francs ($2,300).

In addition to the single-seat versions, Blériot offered two-place variants of the Type XI as well. The most popular of these was the Blériot XI-2. Also known as the "Tandem," the craft could be powered by a 50-hp Gnôme, although the factory recommended the 70-hp version for anything more serious than putting around the aerodrome.

The pilot of the XI-2 sat in front and the passenger behind. If desired, extra fuel tanks could be installed to replace the passenger.

While the dimensions of the Tandem were about the same as those of a standard Type XI, it was a bit heavier at 335 kg. With the 70-hp Gnôme,

A Blériot XI-2 takes to the air.

The Blériot XI-2.

A Blériot XI-2 on the floats. (See Appendix C for details.)

The XI-2 bis "côté-à-côté."

A Blériot XI-2 bis "côté-à-côté."
(See Appendix C for details.)

An XI-2 bis.

the machine could reach 90 kph. Prices averaged some 30,000 francs ($5,800) in 1913.

The Blériot XI-2 was a quite popular airplane. In March 1913 a total of fifty of these machines were either in service or on order by the French Army. An additional thirty machines had been ordered by foreign armies and fifteen had been delivered to civilian aviators.

The Type XI-2 did of course require some redesign. The fuselage frame was extended by adding an additional bay. The wings in the XI-2 variants were also larger in area, with longer spars and additional ribs. The rear wheel or crossed rattan bows of the XI were usually replaced by a single spring-mounted skid in the XI-2 models.

A number of other experimental XI-2 aircraft were also produced. The seaplane version, for example, featured large floats mounted to the front bedstead, with the rudder extending considerably below the fuselage.

Yet another, but much less popular, two-seat model was the XI-2 bis, côté-à-côté, in which pilot and passenger sat side by side. The craft was only sold with a 70-hp engine and was priced at 35,000 francs ($6,800) in 1912.

A military Type XI.

A Blériot Escadrille.

A civil version of the Type XI-2 Artillerie. Note wing cutouts and "high" rudder.

A few XI-3 three-place models were also produced. This craft featured a particularly heavy undercarriage with six wheels. (For information on other multiplace Type XI variants, see Appendix C.)

MILITARY VARIANTS

Virtually all the machines listed above were produced in military versions. Usually these craft had few characteristics to distinguish them from the civilian models, although factory brochures and contemporary journals mention the fact that the military Type XIs were specially designed for easy breakdown and reassembly.

A Militaire broken down for transport.

The triple-tired Blériot XI-3 was powered by a 140-hp Gnôme.

Another view of the XI-3 in civil markings, Reims, 1911.

Some modifications were made to meet special military needs. For example, the Type Onze single seaters were produced with two slightly different modifications. The Militaire was essentially a standard civilian craft, but the Artillerie featured a new "high" rudder and wing cutouts. Both versions were also produced as two-seat, or XI-2 machines. The two-place Militaire was referred to as the Génie by French military officials, while the artillery-spotting two-seater retained the single-place designation Artillerie.

A military version of the XI-3 had also appeared by 1914. A three-place machine with an increased wing area, the machine was powered by a 140-hp Gnôme, but proved ill-suited to combat.

A final military type, the XI BG (Blériot-Gouen), also known as the Parapluie or Vision Totale, featured a parasol wing and a rudder that could be split to serve as a landing brake.

In spite of the obvious advantages of better visibility, and a slightly increased wing area (without the fuselage cutout), the XI BG, employed by the French and Italian air services as well as the Royal Flying Corps and the Royal Naval Air Service, was not an overwhelming success, and saw little use after 1915.

In addition to selling standard Type XIs to military customers, Blériot produced a string of special military monoplanes derived from the basic cross-Channel pattern. One of these aircraft, the Type XXI, a side-by-side two-place machine larger than the XI-2 bis, was also marketed as a civilian craft for 30,000 francs ($5,900).

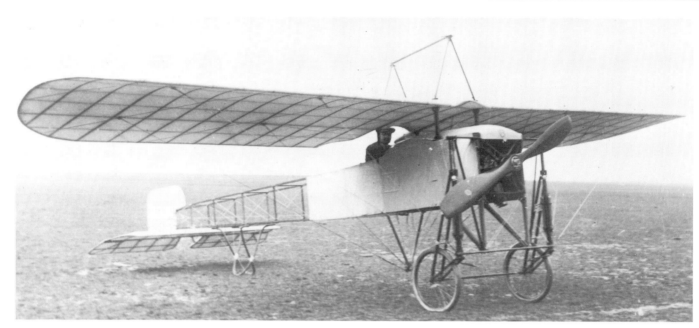

An XI-2 BG, high wing parasol.

A military version of the XI-2 bis. Note the cockpit arrangement.

A Blériot XXI. (See Appendix C for details.)

1911 BLÉRIOT XXI

Span:	11.0 m
Length:	8.24 m
Surface Area:	25.0 m²
Weight:	370 kg
Speed:	90 kph
Tanks:	100 liters (gasoline)
	35 liters (oil)
Duration:	3 hrs.
Engine:	70-hp Gnôme

During the years 1912–14, Blériot produced a number of purely military monoplanes. Often identified as Blinde, or armoured models, they frequently bore little resemblance to the Type XI. (See Appendix C: Blériot XXXVI, 43, and 45, for example.)

RACING AIRCRAFT

The standard and uprated Blériot XI ruled the European racing circuit in 1910. By 1911, however, the firm was producing specially designed racing aircraft, including the Blériot XXIII and XXVII. With completely enclosed, slimly tapered fuselages, specially designed wings, and a 70–80-hp Gnôme, these machines were said to be capable of reaching 120 kph.

In order to match the higher speeds achieved by Nieuport, Deperdussin, and other aircraft after 1911, Blériot resorted to tricks that occasionally proved dangerous. Wing camber became flatter, the span shorter, the structure lighter and weaker, the engine heavier and more powerful. By 1911 some specialized Blériot racers like the Type XXIII had such a high wing loading that they became deathtraps in the event of engine failure. C. G. Grey aptly described the Blériot XXIII with its tiny wing and boxy fuselage as "more like the latter half of a dogfish with a couple of visiting cards stuck on it than anything else."[5]

The craft was first flown at Reims in 1911 by Léon Morane, setting new records for 5, 10, and 20 km. Alfred Leblac and Gustav Hamel both flew Type XXIII aircraft in the Gordon Bennett race at Eastchurch, England,

A Blériot XXVII.

A Blériot XXIII racer, Eastchurch, 1911.

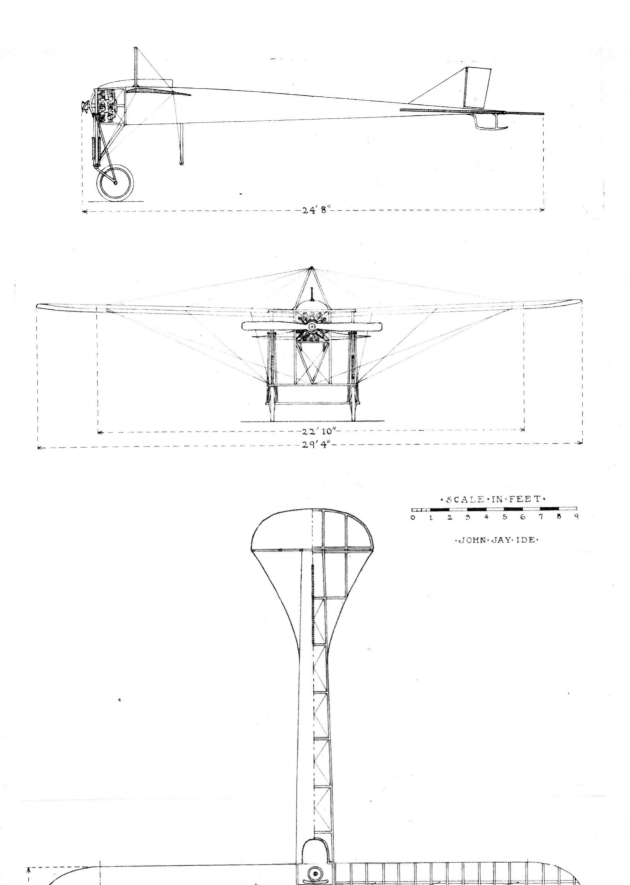

24' 8"

22' 10"

29' 4"

3' 6"

6"

The Blériot XXIII, the "dogfish with visiting cards."

*The Blériot stand at the Third Paris Aero Salon, December 1911. The
XXIII is in the center. To the left is an XI and to the right an XI-2 bis.*

that summer. When the Nieuports entered in the contest proved faster in preliminary trials than the XXIII, Blériot himself simply cut eighteen inches from the already dangerously short wings of both aircraft.

When test-flown with a light load prior to the race, the machines seemed to respond well. Fully loaded on the day of the competition, the situation was quite different. Banking into the first tight turn, Hamel's craft sideslipped to earth. Miraculously, the pilot was thrown clear of the wreckage.

Leblanc was more cautious, making wide turns and remaining in the air to take second place behind a Nieuport flown by Charles Weymann.

But Leblanc had been fortunate. Never again would a Blériot fly with such severely clipped wings.

Other special racing aircraft powered by engines of from 100 to 140 hp were also being marketed by Blériot Aéronautique by 1912. In addition, special factory modifications were available, including alternative engines like the water-cooled Gyp or ENV, a heavier structure to take increased flight loads, and special pilot harnesses designed for stunt flying.

In addition to these Blériot-built originals, the Type XI was perhaps the most frequently copied airplane of the period. Companies in Europe and America produced a bewildering variety of monoplanes based on the Blériot pattern.

The process had begun as early as October 1909, when *Flight* announced plans for the formation of a company to acquire the English rights to build Blériot aircraft. Blériot himself was to serve on the board of the new company, which would also include members of Blériot, Ltd., the firm which marketed Blériot headlamps in Great Britain.

This businesslike licensing arrangement seems to have been more the exception than the rule. A licensee paid a large fee for the right to use Blériot drawings, patterns, and expertise. It was far simpler for a prospective manufacturer to copy, or attempt to improve upon, the basic Blériot pattern without benefit of a license. Thus, officially and unofficially, the Blériot XI design was reproduced in Europe and America.

The experience of the infant United States aeronautical industry with the Type Onze was particularly interesting. The nation that had given birth to the Wright brothers had done little to maintain leadership in aviation. While the Wright and Curtiss companies were producing airplanes in significant numbers, they had by 1910 begun to fall behind the rapid advance of European aeronautical technology. As the editor of a leading American aero journal commented in 1911:

> There are very few cases on record of practical airplanes being built [in America] on entirely new lines. The three score of different makes of aeroplanes today are only further developments or modifications of the half-dozen types that flew in 1909-1910. In many cases the modifications are very slight indeed, so much that it makes one feel that he is abetting in a plagiarism to call them by any other than their original name.[6]

Given this American penchant for "borrowing" basic designs, it was only natural that the interest of U.S. manufacturers and homebuilders would focus on the Blériot XI.

Americans, who had been as excited by the Channel flight as Europeans, had their first opportunity to inspect a Blériot at close range in November 1909. Rodman Wanamaker, the Philadelphia department-store magnate, had purchased the craft while touring France late that summer. Wanamaker had apparently intended to fly the machine, which bore the factory serial number 153, himself, but when detained in Paris on business he shipped the monoplane back to Philadelphia for exhibition and sale at Wanamaker's.

John Moisant, Belmont Park, 1910.

Wanamaker had paid $2,200 for the craft in France, but priced it at $5,000 in Philadelphia. It was purchased by Louis Bergdoll, a wealthy automobile enthusiast, who eventually turned it over to his brother, Grover Cleveland Bergdoll, an early Wright pupil. This original Wanamaker Blériot is now in the hands of Cole Palen, of the Old Rhinebeck Aerodrome in New York.

By the spring of 1911 a number of other Americans, including Earle Ovington and C. J. Stroble, had purchased Blériots for exhibition flying in the United States. Other enterprising souls like Stanley Yale Beach had built their own copies.

Americans living in Europe had also made a number of well-publicized flights on Blériot machines. In August 1910, John B. Moisant, who like Ovington had learned to fly at Pau, became the first man to fly the Channel with a passenger. Two years later, in 1912, Harriet Quimby, drama editor of *Leslie's Weekly*, would become the first woman to fly the Channel. Both Moisant and Quimby would die in Blériot crashes.

In October 1910, John Moisant, Philadelphia's J. Armstrong Drexel, Alfred Leblanc, and England's Claude Grahame-White all entered Blériots in the Belmont Air Meet held near New York. The highlight of the meet was a race between Moisant and Grahame-White, both piloting Type XIs, from the flying field to the Statue of Liberty and back, on October 30. Though Moisant was initially declared the winner, Grahame-White protested that the American had violated the rules of the contest, and the decision was eventually reversed.

The Moisant International Aviators.

In the wake of the Belmont Meet, Moisant and a number of other competitors banded together to form Moisant's International Aviators and took their aircraft on tour, giving thousands of spectators throughout Latin and South America an opportunity to see a Blériot XI in the air. Moisant, who had become one of the best-known American Blériot pilots, died in a crash at New Orleans on December 31, 1910.

The Blériot XI even starred on Broadway during the winter of 1910. James Montgomery's play *The Aviator* (the "merriest kind of melodrama," according to one enthralled critic) featured a full-scale Blériot in which actor Wallace Eddinger was whirled about the stage of the Astor Theater, suspended on piano wires.

In view of the popularity of the Blériot, it was only a matter of time before an enterprising American manufacturer began to build and sell the craft commercially. The Queen Aeroplane Co. of New York was the first to do so in late 1910.

Founded by Willis McCornick, a wealthy stockbroker and sportsman, Queen was originally formed to take over the factory and business of the Lovelace-Thomas Aeroplane and Motor Works, located at St. George Park, 197th Street and Amsterdam Avenue. The plant consisted of an odd assortment of buildings that had once been part of an amusement park. The main assembly building was a skating rink, while nearby structures had been refitted as woodworking and machine shops. In the fall of 1911, Queen employed eighty-five men at the factory and flying field.

Queen Blériots were very close to the original French-built Type XIs in design, though minor modifications of the landing gear and rudder configuration were apparent. In general, they were simpler in construction. For example, brackets were substituted for the standard bell-shaped cloche that controlled the warping and elevator cables of the original (see Chapter V).

Queen Blériots were powered by a variety of engines, but 3-cylinder Anzanis and 50-hp Gnômes were standard. Factory brochures informed potential buyers that Queen had contracted with the Crane Motor Works of Bayonne, New Jersey, for a special "twenty-four-hour non-stop motor, guaranteed to hit like an automobile engine," but it is uncertain whether any of these engines were ever installed.

Indian Rotary engines manufactured by the Hendee Co. were flown on a number of Queen Blériots, including the famous machine used by Earle Ovington to carry the U.S. mail between Garden City, New York, and Mineola, on Long Island, as part of the Nassau Boulevard Flying Meet, September 23 to October 2, 1911. Like the Gnôme, the Indian was a 7-cylinder rotary capable of reaching 50 hp. Special access panels were provided on the Queens to ease the problems of engine maintenance.

These first American Blériot copies came with a well-equipped instrument panel, including a barograph, revolution counter, clock, and angle-of-incidence indicator. Gas and air levers were mounted on the stick. The magneto spark was fixed.

The standard Silver Queen could be purchased with an Anzani for $2,900 in 1911. The Gnôme- and Indian-powered models sold for $5,000 and $4,500 respectively.

There is no record of the number of Queen Blériots produced, although a number of leading American pilots including Ovington, Arthur Stone, and Ladis Lewkowicz flew the machines in meets and exhibitions. Ovington, who had trained at Pau and much preferred the Blériot factory machines, was particularly critical of the Queen monoplanes he flew.

When W. R. Hearst offered $50,000 for the first transcontinental flight

The Queen Blériot.

THE QUEEN MONOPLANE

Earle Ovington carries the mail at the Nassau Boulevard Meet, 1911.

in 1911, Ovington, anxious to compete, was unable to acquire a sufficient stock of spare parts from France and accepted McCornick's offer of a free airplane and enough spare parts for two more. Hendee donated engines for the venture. As Ovington's wife recalled: "I am sure he [Ovington] had more than one regret himself when the American imitation appeared. It lacked the fine workmanship and finish of the French machine. A casual observer might not be able to tell it from the Dragonfly [Blériot XI], but anyone who knew the least thing about airplanes—even an aviator's wife—could see the difference at once."[7]

Ovington himself remarked that American manufacturers like Queen " 'improve' the planes which they try to imitate until they won't leave the ground, and if they do go up they are poor flyers." After repeated alterations to the Queen, and so many false starts that the venture became a joke in the press, Ovington finally abandoned his dream of crossing the continent. His wife, Adelaide, recalled his parting comment on the Queen Blériot: "Well, I've done all I can to make this combination fly. But the Lord never meant it to stay in the air, and far be it from me to dispute the matter any longer."[8]

rst successful two-seated
leriot type Monoplane flown
in the U.S.A.

— Built by —
The American Aeroplane
Supply Ho
Hempste

Our type of
single-seater machine
purchased and
flown by
Willie Haupt.

An American Aeroplane Supply House Blériot XI-2.

1911 SILVER QUEEN MONOPLANE

Span:	8.84 m
Length:	26.0 m
Weight:	355 kgs (empty)
Speed:	90 kph
Tanks:	27 gals. (gas); 13 gals. (castor oil), gravity feed
Engine:	Anzani; Gnôme; Indian

Another New York firm, the American Aeroplane Supply House, founded by Frederick C. Hild and Edward F. Marshonet, also produced Blériot types. Established as a pure supply house specializing in imported Blériot parts, the firm sold its first full-scale airplane, a Type XI, to William E. Haupt of Philadelphia in the spring of 1911. Haupt, who had been a race-car driver for the Bergdoll Motor Co., had taught himself to fly on the Wanamaker-Bergdoll Type XI. He had then moved to Mineola, New York, where he flew a number of other Blériots before ordering his own machine from the American Aeroplane Supply House (AASH).

According to stories circulated at the time, Hild based the Haupt machine on measurements he had made of a Type XI that Earle Ovington had brought back from France in March 1911. Judging from photos of the Haupt craft, it was a fairly faithful copy of an uprated Type XI, with the exception of the engine, a 4-cylinder, 50-hp Roberts water-cooled in-line.

After successful test flights of his new machine at Mineola, Haupt took part in a number of exhibition dates in Pennsylvania and New Jersey, then returned to New York to become a test pilot for AASH. By late July 1911, when Haupt entered AASH employ, the firm was expanding its offerings to include a variety of Blériot types. Between 1911 and 1913 four basic models were produced: a single-seater, a passenger version, a racer, and a military model. The standard Blériot XI single-seater was available with engine choices ranging from imported 3-cylinder Anzanis and 50–70-hp Gnômes to water-cooled 4- and 6-cylinder, 50–75-hp Roberts engines of U.S. manufacture. Prices ranged from the Anzani model at $2,700 to the 70-hp Gnôme version at $6,000.

Engine choice for the two-place machine was similar, with a 6-cylinder 60-hp Anzani being substituted for the 3-cylinder version. Engines available for the racing monoplane included 100- and 140-hp Gnômes costing up to $10,000.

Once again, precise production figures for AASH Blériots are unavailable, but the numbers were very small. *Jane's* lists only six aircraft produced in 1911. Factory sales brochures repeatedly referred to six "prominent owners of our Blériot type monoplanes": Haupt; A. C. Menges of Memphis, Tennessee; E. J. Morley of Sumner, Mississippi; J. Albert Brackett of Boston; A. V. Reyburn of St. Louis; and Charles W. Spencer of Philadelphia.

In addition to the finished Type XI copies produced by firms like Queen and the AASH, amateur aircraft builders across America tried their hand at constructing Blériot reproductions. Aeronautical journals and the publishers of "how-to" books encouraged these efforts, providing detailed descriptions, drawings, and step-by-step instructions as well as passing on advice from other homebuilders. The appearance of Lawrence L. Prince's multipart "How to Build a Blériot-Type Monoplane" in *Aero* beginning in April 1911, and the publication of Charles B. Hayward's *Building and Flying An Aeroplane* in 1912, were particularly influential. The existence of supply houses like the AASH and E. O. Rubble of Louisville, Kentucky, which specialized in providing metal fittings and imported parts for homebuilders, also encouraged the construction of backyard Blériots.

A number of these American home-built Type XIs survive in museum collections. The most notable are two Blériots constructed by Ernest C. Hall of Warren, Ohio, now on display in the U.S. Air Force Museum, Dayton, Ohio, and in the Bradley Air Museum, Windsor Locks, Connecticut.

By 1913 the Blériot XI had become an obsolete design. While scattered amateurs in the United States and Europe continued to build copies and Blériot Aéronautique was still producing variants as trainers (and the high-wing Vision Totale that saw observation duty early in World War I), the pace of aeronautical progress had moved far beyond the classic cross-Channel design. But its influence had been unmistakable. Not only was it *the* monoplane of the prewar era, mass-produced and copied in unrivaled numbers, but it had given birth to a family of monoplane designs. The debt owed to the little short-span monoplane by firms like Morane-Saulnier, Blackburn, Fokker, and Nieuport scarcely requires comment. After the Wright Flyer, the Blériot XI was clearly the most influential craft of the period.

But in spite of its enormous popularity and influence, the Type Onze was heir to a variety of aerodynamic and structural woes that would lead many engineers and military officials to question its basic airworthiness.

Léon Delagrange just before taking off for his last flight.

IV

The Problem with Monoplanes

At about 2:30 on the afternoon of January 4, 1910, Léon Delagrange was completing his third circuit of the aerodrome at Croix d'Hins, near Bordeaux. As he eased into a turn at an altitude of some forty feet the left wing of his Blériot suddenly folded straight down, sending plane and pilot crashing to earth. Delagrange died instantly, the first victim of a monoplane crash.

Naturally the accident attracted wide interest. In spite of the obvious dangers faced by aviators, Delagrange was only the fifth powered-airplane fatality.

The popular presumption was that pilot error had been responsible for the crash. Blériot himself pointed out that Delagrange had been flying the first Type XI fitted with a 50-hp Gnôme that raised its speed from 40 to 50 mph. He argued that the increased speed, the added weight, and the gyroscopic effect of the spinning engine may simply have been too much for the pilot to handle.

Other observers thought this unlikely. Delagrange had flown the machine before with no apparent difficulty and was one of the most experienced pilots in Europe. Like Voisin, Levavasseur, and Henry Farman, he had attended the Beaux Arts and had built an enviable reputation as a sculptor before becoming one of the first successful pilots of the classic Voisin biplane. He won the Aéro-Club de France 200-meter distance prize on March 28, 1908, and had engaged in a friendly competition with Farman throughout the spring of 1908 that led to repeated extensions of the European distance record.

Fascinated by the Wrights, he had learned to fly one of their machines. Then, in September 1909, he became the third man (after Blériot and Leblanc) to fly a Type XI.

Hubert Leblon, an ex-motorcycle and -automobile racer and an aviator himself, suggested a more frightening possibility: structural failure. Leblon, who served as a mechanical advisor to Delagrange, recalled that the pilot was "perpetually anxious when (whilst flying at high speeds) he heard the supporting wires hum and quiver in the wind." Perhaps one of these wires had snapped.[1]

Henry Farman agreed, pointing to the obvious fact that the loss of the flying wires would "inevitably cause the destruction of the monoplane."[2] As H. R. D'Erlanger remarked in a letter to the editor of *Flight:* "If this is so, the accident is not due to a defect of that particular monoplane, but to an inherent defect in all monoplanes . . . and I think the question it raises is one deserving of attention from designers." Over the next three years it would indeed receive a great deal of attention from designers as the toll of monoplane deaths mounted.[3]

Ironically, Leblon himself was the next man to die in a Blériot. Like Delagrange, he was flying a Gnôme-powered Type XI. On April 2, 1910, having been honored at a luncheon given by the citizens of San Sebastian, Spain, he took off for a flight over the sea. On his return, the aircraft dived into the rocks lying just beneath the surface of the water. Neither engine

Georges Chavez is held back for take-off.

failure nor a sudden illness of the pilot, both of which were advanced as potential causes, seemed to explain the tragedy. Several witnesses later recalled that the wings of the craft had folded just before the crash.

That fall, a far more celebrated crash focused attention on the safety of the Blériot once again. On September 23, 1910, George Chavez, a young Peruvian pilot, took off from a small field near Brigue, Switzerland, in a Gnôme Blériot XI. His goal was to cross the Simplon Pass (elevation 2,100 meters) and fly seventy-five miles over the Lombardy Plain to Milan, to win a 70,000-lire prize offered by the organizers of an Italian flying meet.

"La traversata delle Alpi in aeroplane" would be an extraordinarily difficult and dangerous flight. Treacherous mountain winds, the uncertainty of engine operation in the high thin air, and the absence of any landing fields prior to Simplon Pass led the organizers of the meet to limit entry in the contest to the most experienced aviators. By the time the flight began, the field had been narrowed to two men, Chavez and the American Charles Weymann. Chavez had attempted the flight on September 19 but was turned back by high winds. Weymann made three tries on the 21st but was forced back each time by carburetor icing.

Chavez took off for the final try at 1:29 P.M. on the 23rd. Forty-two minutes later he was through the pass and descending toward the village of Domodossola. Only thirty feet up, he blipped his engine, apparently to lengthen his glide. Seconds later his machine was a tangled wreck. Chavez died two days later in the hospital at Domodossola.

In the immediate aftermath of the accident, pilot error was once again a favorite explanation. The intense cold, argued many, had so debilitated Chavez that he was unable to operate the controls. But as a few eyewitness accounts of the crash began to make their way into print, it became obvious that, as in the case of Delagrange and Leblon, the wings of the Chavez Blériot had collapsed in flight.

Over the next two years, the problem of monoplane wing failure became ever more apparent. On the surface, the statistics were not too frightening. An admittedly incomplete count shows that between 1908 and 1912, some

113 men and women lost their lives in aircraft crashes. Fifty of these had been in monoplanes. Considering the fact that Claude Grahame-White had counted 361 biplanes and 302 monoplanes in service in 1910, and assuming that this proportion did not change drastically over the next two years, it will be seen that the number of monoplane fatalities was not disproportionate. Even the fact that more people had died in Blériots than in any other aircraft type was not too disconcerting, considering how many more Blériots were flying. According to Grahame-White's figures, the Type XI was the single most popular airplane by 1910.[4]

But other factors were involved. Most accidents with biplanes could honestly be traced to pilot error or to the structural failure of some part of a particular airplane. It was universally agreed that the biplane was a more satisfactory solution to the problem of aircraft structural design than the monoplane. Since 1896, when Octave Chanute and A. M. Herring had produced their classic strut-and-wire-braced "two-surface" glider on the basis of experience with bridge trusses, biplane construction had been recognized as stronger, safer, and lighter than monoplane construction.

Engineers like Blériot had been willing to pay a penalty in weight, higher

The wreck of a Blériot flown by Léon Morane.

wing loading, higher landing speeds, and the like because the monoplane offered the advantages of decreased resistance and the absence of the aerodynamic interference that occurred between the wings of a biplane. This meant that a monoplane would generally be faster than a biplane powered by the same engine. To a pilot or manufacturer seeking publicity, the heavier wing construction and wire bracing required to support the flight loads on cantilevered monoplane wings seemed a small price to pay for even a few additional miles per hour that might mean the difference between victory and defeat in the long-distance and closed-course air races of the period.

But the increasing string of monoplane wing failures led many engineers to conclude that there remained some as-yet-undertermined flaw inherent in all single-wing airplanes. As Granville E. Bradshaw of the Royal Aircraft Factory remarked in 1912: "There seems to be a feeling amongst us all that monoplanes as at present constructed are not everything to be desired from a mechanically sound point of view, and indeed the more one investigates the loading on wing wires and main spars the more does one feel concerned about machines of this type that are flying."[5]

As the most popular monoplane, the Blériot XI was an object of particular concern. As early as December 1910, engineer John H. Ledeboer had noted that "there is too a small margin of strength in the construction of some points of the framework of the [wings] . . . Two, at any rate, of the fatal accidents that have occurred in the flying of Blériot monoplanes are traceable to the fracture of the plane framework in mid-air . . . The problem is to find the weak point in the structure."[6]

Removing the wreckage following a Blériot crash that took the life of French Army Lt. Blanchard.

James Radley flying his Blériot monoplane.

Concern over the structural integrity of the monoplane reached a climax during the spring and summer of 1912. As a result of three accidents occurring in rapid succession, the French were the first to take action. On December 13, 1911, Lieutenant Lantheaume, an experienced military pilot, died in the crash of his Blériot following a cross-country flight to Milan. Lt. Robert Ducourneau was flying near Pau at an altitude of 150 meters on February 22 when the wings were torn from his Blériot. Lt. Henri-Paul Sevelle was the next victim. Circling the military aerodrome at Pau on March 13, his machine broke up in the air and crashed.

The engineers at Blériot Aéronautique had not been idle as the number of suspicious crashes mounted. Following Chavez's death they had strengthened the wing structure of the Type XI. The wingspars had been reinforced a second time after the Leblon crash.

By the time of Lantheaume's death, French military officials were beginning to take note of the situation. A special commission was directed to study the monoplane problem and recommended that the wings be reinforced once again.

Throughout this period Blériot was conducting a rigorous test program to insure that the wings and flying wires of his monoplanes were strong enough to support expected flight loads with a reasonable margin of safety. The standard test procedure called for turning an aircraft on its back and spreading a quantity of sand weighing four to six times the expected normal flight load on the underside of the wing. In the absence of any serious understanding of the nature and the magnitude of the stresses encountered in flight, or any agreement as to what constituted a reasonable factor of safety for a high-performance aircraft, sand testing remained a primitive means of gauging the strength of a structure, yet it was standard practice, and the Blériot machines repeatedly passed with flying colors.

The problem apparently did not lie in the ability of the wings to carry normal vertical loads. The search was on for a dangerous stress coming from a direction engineers had not considered.

R. F. Macfie, an Englishman, suggested such an alternative in the aftermath of the Delagrange crash. Macfie had long been concerned about the stresses imposed on monoplane wings by simple head resistance. He believed that he had found evidence to support his theory in the remains of the Delagrange machine, in which both front and rear left wingspars had broken, while the flying wires were unbroken. A second accident to James Radley at Huntingdon on May 10, 1910, provided more evidence for Macfie. Radley had been flying at an altitude of fifty feet when his machine suddenly swung to the right. After struggling to land, the pilot discovered that both right spars were cracked for about two feet of their length beginning at a point some two feet from the fuselage. The rear spar was not only cracked but severely twisted, and the entire wing was bent back at an alarming angle. Macfie reasoned that the combination of head resistance, skin friction on the wings, and the high speeds made possible by the new Gnôme engines were sufficient to explain the Delegrange and Radley crashes.[7]

Contemporary engineers realized that Macfie's analysis was faulty. Head resistance and skin friction were simply inadequate to explain the problem. But they had a much more difficult time ignoring Louis Blériot.

Like Macfie, Blériot provided a chatty discussion of the problem, devoid of any mathematical treatment and depending on common-sense reasoning. Blériot released his report on the problem in the spring of 1912, as the string of monoplane accidents was leading French military officials to consider seriously the wisdom of future monoplane purchases. He believed that the cause of the wing failures was to be found in top-loading. A monoplane had two sets of bracewires. Those beneath the wings, the flying wires, helped carry the normal loads encountered in flight and were much heavier than the landing wires which guyed the wings from above and supported their weight when the machine was on the ground.

Blériot pointed out that his program of sand-testing had demonstrated his monoplane's ability to support normal flight loads, but argued that an aircraft pulling out of a steep dive would encounter forces pushing down on the wings from above. "There is no longer any room for doubt," he remarked, "that the deaths of Chavez, Blanchard, and Lantheaume were caused, not, as has been believed up to the present, by the breaking of the wings, that have passed their trials and tests of positive loading successfully, but by the failure of the upper guys which have no strength to resist these forces coming from above." As evidence for his contention, Blériot cited the reports of witnesses who claimed that the wings of the Delagrange machine folded down, not up, and pointed to the fact that the landing wires of the Ducourneau, Lantheaume, and Sevelle machines were broken, while the flying wires were intact.[8]

Blériot was widely praised for facing the problem squarely. The English journal *Flight* was typical, remarking: "That it should have been written by the French pioneer designer of monoplanes and should form such a frank and lucid exposé of a hitherto unsuspected weakness in such machines, is the finest possible vindication of the Etablissement Blériot as a scientific concern."[9]

The response to Blériot's report from the engineering community was polite, but few technicians were willing to accept his suggestion as a complete answer to the monoplane problem. Leading aircraft designers, including A. V. Roe, L. Howard Flanders, and Robert Blackburn, found it difficult, in Roe's words, "to conceive a pilot making a sufficiently sudden descent to cause excessive down thrust on the wings."[10] In any case, if Blériot was correct, the solution was simple: strengthen the bracing.

Others, notably Mervyn O'Gorman, superintendent of the Royal Aircraft Factory, took a less positive view, arguing that while Blériot's presentation showed great courage, it was so chatty and imprecise as to be of little value to engineers in search of the cause of monoplane wing collapse.[11]

The French government was satisfied with Blériot's explanation, however. In the wake of his report, all monoplanes in military service were temporarily grounded until both the flying and landing wires were strengthened.

Douglas Graham Gilmour.

In addition, a strenuous testing program designed to investigate the structural integrity of the Blériot was undertaken in May and June of 1912. Directed by Lieutenant Colonel Etienne of the Technical Department of the Vincennes Military Aviation Establishment, the tests were witnessed by Colonel Hirschauer, Permanent Inspector of Military Aeronautics, and Colonel Bauttieaux, Director of Military Aeronautics at Chalais Meudon. During the most vigorous phase of the trials, a standard Blériot XI was mounted on a flatcar drawn by a locomotive provided by the Compagnie du Nord. The train was run back and forth along a 5-km stretch of track near Chantilly at speeds of up to 115 kph, some 10–12 kph in excess of the top airspeed of the machine. Captain Charet and Lieutenant Maillot took turns in the cockpit, forcing the monoplane "to assume the different positions of ascent, descent, and warping as quickly and roughly as it could possibly be done in an endeavor to realize the very worst conditions which the machine might have to fight." Once again, "all parts of the Blériot machine stood the test perfectly, as was afterwards testified by the military officers present."[12]

With French confidence in the Blériot restored, concern over the failure of monoplane wings shifted to England, where the crisis reached a peak in the summer and fall of 1912. The death of Douglas Graham Gilmour in a Martin-Handasyde monoplane on February 17, 1912, had shocked the English public. Young, handsome, the most popular of English aviators, Gilmour was passing over the Old Deer Park at Richmond, not far from London, when his right wing was seen to give in the middle. The craft swung around as Gilmour struggled to maintain control. A moment later the left wing cracked and the machine dropped straight to earth.

An examination of the wreckage only added to the confusion. None of the bracewires were broken; the control wires were intact; and all controls seemed to be in working order. As C. G. Grey of *The Aeroplane* commented: "What was the cause of the accident will . . . never be known, and one can only surmise that there must have been some undiscovered flaw in the wood of one wing, although this seems impossible, seeing the size of the spars and the care with which Martin and Handasyde select their material. Further, the king posts of the wings have an enormous factor of safety, and the wing spars alone are strong enough to stand the load without the stays from the king-posts outwards. The whole thing is an inexplicable mystery." The reaction was immediate. In the words of one commentator: "Poor Mr. Gilmour's fatal accident has called forth from the ignorant public an outcry that aviation should be stopped, that aeroplanes can never become safe, that all aviators must be mad, and so forth."[13]

Nor was this attitude so difficult to understand. In spite of the best efforts of Louis Blériot and other leading engineers, the collapse of monoplane wings remained as mysterious as ever. One puzzled designer put the problem most succinctly: "We know a wing broke, but we do not know how or why it broke."[14]

The situation rapidly grew worse following Gilmour's death. Monoplanes were responsible for every fatal aircraft accident that occurred in Great Britain over the following months. Military officials were particularly worried. Twenty-one of the approximately fifty-six aircraft in the recent British inventory were single-wing machines, a mix of Blériots, Nieuports, Bristols, Deperdussins, Shorts, and Etrichs. The generally dismal record of the monoplane, coupled with several serious monoplane crashes suffered at the Military Trials that year and the short-term French grounding of monoplanes, led English officials to consider a similar ban to allow time to investigate the problem.

Word of the proposal to ground military monoplanes leaked to the aeronautical press in late September 1912. Knowledgeable commentators reacted with alarm, hoping that "the decision of the authorities temporarily to prohibit the use of monoplanes on army service [would not be] wrongly interpreted by the man on the street." Yet there was a widespread recognition of the need for an investigation that would either pinpoint the problem or demonstrate that there were no inherent flaws in the monoplane as a type.[15]

A committee "to enquire into and report upon the causes of the recent accidents to monoplanes of the Royal Flying Corps and upon the steps, if any, that should be taken to minimize the risk of flying this class of aeroplane," was named early in October. During the course of the investigation the nine-member panel (a mix of pilots, military officials, and such leading engineers as O'Gorman and F. W. Lanchester) drew heavily on the resources of the National Physical Laboratory in studying the causes of the accidents. The reports of the Public Safety and Accidents Investigation Committee of the Royal Aero Club also proved to be of value.

The group issued its final report in April 1913. As aeronautical enthusiasts had hoped, the committee determined that the three specific accidents on which they had concentrated "were not primarily due to causes dependent on the fact that the machines were monoplanes." Specific recommendations focused on improved maintenance and inspection procedures, but, as in France, the monoplane had been absolved of inherent defects.[16]

Yet in spite of official approval in France and England, the problem of the monoplane remained. As late as January 1913, Griffith Brewer, an English aeronautical authority and friend of the Wright brothers, could still remark that it was "virtually necessary either to abandon the use of monoplanes altogether or to look more deeply in order to ascertain the cause of this type of structural collapse."[17]

But the "cause of this type of structural collapse" would remain hidden from the engineers of the prewar era. Some designers, like Blériot, solved the mysterious difficulty empirically, strengthening the wingspars and wires of their aircraft repeatedly until the problem simply disappeared.

Not until the years after World War I would a more complete understanding of the complex forces acting on a wing begin to emerge, enabling engineers to analyze aircraft structures and avoid the unseen dangers that had plagued their predecessors of the 1909-18 period.

In spite of the relatively primitive analytical techniques at their disposal, a few of Blériot's contemporaries were able to suggest the existence of unsuspected stresses that might lead to the failure of a wing that had demonstrated its ability to carry up to six times the expected vertical flight load in sand tests. Griffith Brewer, for example, believed that torsional or twisting forces were at work. While he could do little more than speculate, Brewer described a condition that aerodynamicists now recognize as wing torsional divergence.

Every beamlike structure, including a wing, has a shear center around which it will twist unless the load is applied directly at that point. The braced wings of a biplane form a torque tube of large cross-section that effectively resists this twisting action. Monoplane wings, in which the torque tube consists only of the cross-section of the single wing, are much more susceptible to torsion. Thus, if the center of pressure on a monoplane wing does not fall on the shear center, the wing tips will not only be lifted but twisted to the front or rear depending on the location of the shear center. This torsion increases the wing loading, which further increases the twist and so on until the aircraft reaches its divergence speed, at which point the wings are twisted from the fuselage.

With the exception of occasional hints like Brewer's, the possibility of torsional divergence was not suspected until the 1920s. Two Swiss engineers, A. Eggenschwyer and A. Maillert, published their work on the shear center in beams in 1920–21, while aerodynamicist Hans Reissner presented his theory of wing-load distribution, divergence, and other aeroelastic phenomena in 1926.

Modern engineers, operating with benefit of hindsight, have identified torsional stress as a problem leading to the failure of several early monoplane wings, including that of the Langley Aerodrome of 1903. The problem of torsional divergence may also have been a contributing factor in the much-publicized failure of the Fokker D. VIII wing late in World War I, although poor workmanship and shoddy materials were also problems in this case. It is quite possible that it was also the cause of some of the much-publicized monoplane crashes of the period between 1910 and 1913. It was not, however, the problem in the case of the Blériot XI.

Modern analytical methods reveal the divergence speed of the cross-Channel Blériot XI to be in excess of 100 mph. This is far beyond the top speed of even the much-improved Type Onze of 1914, which, with its stiffer wing and a shear center located considerably closer to the center of pressure, had a much higher divergence speed.

But torsion was not the only unsuspected force operating on the frail wings of prewar monoplanes. As early as 1911, P. James, a French mining engineer, had called attention to the complex stresses imposed on wingspars by the very guywires that were intended to brace them.[18]

The wings of the first Blériot aircraft were composed of two main spars of equal dimension. Each spar was a beam, a structure ideally suited to carry the transverse loads imposed by lifting pressure, but not to carry axial, or compression, loads.

Few engineers of the period considered the danger of such compression loads operating on a spar. Yet as the speed of an aircraft increased, the lower flying wires, designed to help hold the cantilevered spar in place, imposed just such loads. As the speed and wing loading increased, the spar carried not only an increasing vertical or transverse pressure, but axial or compressive forces caused by the bracing as well.

In the case of the original Blériot, the situation was complicated by the basic wing design. Because the two main wingspars were of equal dimension, the shear center was located exactly between the two beams, just forward of midchord, while the center of pressure fell close to quarter chord. Thus, the wing did have a tendency to twist, though not sufficiently to cause failure. Nevertheless, these twisting stresses on the outer fibers of the spar exacerbated the compressive strain imposed by the guywires.

To add to these difficulties, the fact that the center of pressure was at quarter chord meant that the forward spar would bear nearly two-thirds of the normal flight load.

All of this suggests beam-column failure as the most probable explanation for the collapse of early Blériot wings. A recent analysis of the wing of the cross-Channel Blériot XI undertaken by Dr. Howard Wolko of the National Air and Space Museum indicates that such a failure might be expected with the aircraft traveling near 50 mph. It might be initiated by any sudden increase in wing loading as would be encountered when pulling out of even a gentle dive, or adding power after a gliding descent.[19]

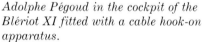

Adolphe Pégoud in the cockpit of the Blériot XI fitted with a cable hook-on apparatus.

Dr. Wolko's analysis suggests an initial failure of the front spar two to three feet out from the fuselage, followed by the failure of the upper, or landing, wires, as the broken wing was snapped down by the pressure of the flying wires.

This explanation, which fits the Delagrange, Chavez, Radley, and Gilmour crashes, would have been as puzzling to Blériot as torsional divergence. Muller-Breslau had published the basic formulas for beams subjected to combined loads in 1902, but his initial work was so difficult to apply in practice that it was virtually ignored by engineers. By 1913 the English engineer L. Bairstow had developed simplified methods of calculating combined stresses, but these were less accurate than the Muller-Breslau formulas and did not yield fully trustworthy conservative figures. During World War I, Arthur Berry derived the tables of complex functions that made the earlier work useful, but even so the formulas were not commonly applied until further refinements were made by staff members of the Army Air Service Engineering Division at McCook Field in 1922.

But in spite of his inability to analyze the condition properly, Blériot had reached an empirical, common-sense solution to the problem by mid-1912. One element of the solution was simply to strengthen the spars and wires. In addition, he came to recognize the fact that the front spar carried a higher load and should be larger than the rear spar. The larger front beam could not only carry a higher axial load in compression but had the additional advantage of moving the shear center of the entire wing closer to the center of pressure, thus reducing torsional stress on the spar. Finally, the addition of smaller intermediate spars, or battens, and the use of an aluminum leading edge helped create a stiffer, stronger wing capable of carrying higher transverse, axial, and torsional loads.

With all of these new elements in place, Blériot was certain that his structure was trustworthy. Yet in spite of the many improvements and the official approval of two European governments, the monoplane remained under a cloud through the spring of 1913.

This deep-seated crisis of confidence in the Blériot was finally overcome in the most dramatic possible fashion when Célestin Adolphe Pégoud astounded Europe with the first aerobatic exhibitions over the Blériot aerodrome at Buc in September 1913.

Pégoud was not the first man to fly a loop. That honor goes to Lt. Petyr Nesterov of the Imperial Russian Air Service, who performed the world's first loop at Kiev in August 1913. But Pégoud was the first man to demonstrate the full aerobatic potential of the airplane.

He had begun flight training early in 1913 at age twenty-four. Blériot himself was so impressed with the rapid progress of his student that Pégoud was hired to work with John Domenjoz and Edmond Perreyon, chief pilots of the firm, within a month of his arrival.

Pégoud quickly surpassed his mentors and was chosen to demonstrate a hook-on cable system that Blériot had devised to enable aircraft to operate from shipboard. But Pégoud's major interest was in flight safety. On August 19, 1913, he made the first test jump with the new, lightweight Bonnet parachute. He realized, however, as did Blériot, that the parachute was only a last resort. The real challenge was to build a safer aircraft and to teach pilots what could and could not be done with their machines.

Pégoud was confident that the Blériot XI was now much safer than most pilots supposed. He was determined to demonstrate that the safe pilot was one who had complete mastery of his machine and would not be thrown into a fright when placed in an unfamiliar position.

He began his campaign on September 1, 1913. Climbing to 3,000 feet over Buc, he performed a vertical S, nosing into an outside half-loop, pulling out inverted, and flying for a few seconds before diving into an inside half-loop to recover. The feat was repeated the following day before military officials.

On September 21, 1913, Pégoud completed his first loop, an event that electrified European aviators who were not aware of Nesterov's performance a month earlier. Blériot had carefully inspected the aircraft, a beefed-up 1911 Type XI with a higher cabane than usual, prior to the flight. With Pégoud strapped into the aircraft in a special harness, Blériot himself spun the propeller. The pilot climbed to 3,000 feet and four times attempted to loop without success. After landing to discuss the situation with Blériot he was set for another try. This time he was successful. After climbing a final time to perform what one English observer referred to as his "old stunt," the vertical S, he landed to receive the accolades of the enormous crowd.

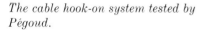

The cable hook-on system tested by Pégoud.

Pégoud "loops-the-loop."

Veteran pilots were aghast. Men who had hesitated even to bank their aircraft had seen Pégoud perform apparently suicidal maneuvers with impunity. The editor of *Flight* spoke for many when he remarked: "Human nature presents itself in many guises, but we venture to think that it does not often evolve a type so utterly full of nerve as Pégoud."[20]

Blériot, of course, was overjoyed. Having just emerged from a period when the basic structural integrity of his aircraft was being questioned, it must have been pleasant indeed to read that "the experiments have proved that the Blériot monoplane must be very seriously and strongly constructed to go through such hard tests, and that no small part of the achievement is due to M. Blériot for the design of such a splendid machine."[21]

Within a few days of accomplishing his first loop, Pégoud was off to Brooklands, in Surrey, to give the English their first lesson in aerobatics. He performed his now standard vertical S's and inside loops and also included tail slides and a half-loop with a roll-out at the bottom during which he momentarily held the wings vertical. Fortunately, even Pégoud did not attempt an outside loop, although early accounts of his activity erroneously suggested that he had accomplished the feat.

By the end of Pégoud's three-day exhibition at Brooklands, most engineers and pilots had come to see the value of aerobatics. Remarked one observer:

> When we first heard of Pégoud and his upside-down flights, we not unnaturally regarded it more in the light of a sensational stunt than anything else. Having seen his actual performance, however, we are of the opinion that it is the most scientific flying exhibition that has even been made . . . his exploits in the air demonstrate the fundamental theory of [good] aeroplane design . . . A well-designed aeroplane, with a tendency to assume a natural head-first gliding attitude . . . is a safe machine so long as it remains intact and the pilot remains on board. Even when brought to a standstill in the air, it will still, in due course, recover in a controllable position.[22]

Blériot and Pégoud had made their point. Back at Buc in early October, Pégoud flew for up to fifty-nine seconds inverted, looped six times in succession, and performed the first loop in a Type XI-2, the rear seat weighted with shot bags.

By mid-October he had indoctrinated Domenjoz and Perreyon into aerobatics, and the three men had formed the first formation stunt-flying team, performing before crowds of up to 200,000 at Buc. Their show was likened to "the movements of a troupe of American step dancers. The machines flew together, rising and planing, and making very sharp turns, following a prearranged programme, all with a rhythmic beauty which was amazing." Pégoud, Domenjoz, and Perreyon had invented the air show. More important, in forcing other pilots to recognize the importance of mastering the full potential of the airplane, they helped set the stage for the much deadlier aerial maneuvering that would take place in the skies over France during the next five years.[23]

By the summer of 1914 Louis Blériot was firmly established as one of Europe's leading aircraft manufacturers, but he no longer held the commanding position he had enjoyed in 1910. None of the designs that he had developed since 1909 had proved as popular, or saleable, as the original Type XI, which was already well on its way to obsolescence.

When Blériot Aéronautique moved into a new factory at Suresnes in 1914, Blériot was seriously casting about for a new project that would enable him to continue expansion. He found it in the Deperdussin, a monoplane racer designed by Louis Bechereau and Fritz Koolhoven that had attained

FLUG a. RÜCKEN d. AEROPLANS.

Pégoud flies upside down before a crowd of German spectators.

a speed of 124.8 mph the previous August. When Armand Deperdussin, a silk merchant who founded the firm manufacturing the racers, was jailed for "business misadventures" in 1913, Bechereau, the technical director, carried on the business in order to clear the debts of the firm.

Soon after the outbreak of war in August 1914, Blériot stepped into the breach, restructuring the Deperdussin company as the Société Pour Aviation et ses Dérives in order to keep Bechereau and his talented colleagues at work. The result was the famous series of SPAD fighters, perhaps the finest "aéroplanes de chasse" to emerge from the war.

From 1914 through 1918, the SPAD and Blériot firms were united at the Suresnes plant. A few original Blériot designs were produced, most of them Type XI derivatives like the parasol wing Vision Totale, but for the most part Blériot's productive capacity was devoted to the manufacture of SPADs and other craft better suited to the rigors of modern aerial combat.

Blériot's tradition of hiring fine designers continued after the war. The firm turned out a variety of novel experimental machines ranging from single-engine monocoque fighters to four-engine flying boats that saw limited service on the South Atlantic mail routes.

But Blériot aircraft were not purchased in quantity by the French military. C. G. Grey commented on the situation at the time of Blériot's death of heart disease in August 1936:

> His friends said that he did not know how to use the right kind of graft, and his enemies said that he did not know how to make the right kind of aeroplane. But on the whole I cannot recall any Blériot type of the past eighteen years which was actually as good as the types which were ordered by the French government. If they had performance they failed in other flying qualities, and if they were nice to fly they were down on performance.[24]

Louis Blériot never recaptured the enormous success of the prewar years when the Type XI dominated European aeronautics.

John Domenjoz (1886–1952).

The Domenjoz Blériot.

Restoring the Domenjoz Blériot

The last of the peacetime Type Onze aircraft were rolling out of the factory at Levallois in July and August 1914. While they had already been superseded by much faster clipped-wing racing monoplanes on the one hand and improved military scouts on the other, Blériots were still being sold in quantity as trainers and remained a popular choice with exhibition pilots.

One of the aircraft produced that July was earmarked for John Domenjoz, a Blériot factory instructor and chief pilot who planned to tour Europe with his new machine. Domenjoz was a native of Switzerland, born in Geneva on April 6, 1886. He had begun flight training at Pau in March 1910 and soloed a Type XI trainer powered by a 35-hp Anzani on May 10, 1910. On February 3, 1911, he earned the 33d Fédération Aéronautique Internationale pilot's brevet to be awarded by the Aero Club of Belgium.

A skilled pilot, Domenjoz was invited to remain at Pau as an instructor after receiving his training. He taught a number of extraordinarily adept pupils over the next two years, including Pégoud. In company with Pégoud and Edmond Perreyon, the other chief pilot for Blériot Aéronautique, Domenjoz earned a reputation as one of Europe's most successful stunt fliers, giving exhibitions in major European cities through the early summer of 1914.

Domenjoz planned to continue his exhibition tour after accepting delivery of his new Blériot in July 1914, but the war clouds descending over the continent forced him to pack his machine and travel to South America, where he continued to thrill crowds by looping the loop, demonstrating dead-leaf falls, corkscrew turns, and other daring maneuvers. During the course of his exhibitions he flew inverted for up to a minute and twenty seconds, a feat that earned him the soubriquet "upside-down Domenjoz."

On September 28, 1915, John Domenjoz and his Blériot XI arrived in New York. G. J. Kluyskens, a United States dealer for Blériot aircraft and Gnôme and Anzani engines, had arranged for Domenjoz to give exhibition flights at the Sheepshead Bay racetrack on Long Island, where he quickly became a major drawing card.

His flights over New York City also drew attention and comment. On October 13, 1915, while he was attempting to circle Manhattan, his engine quit and he was forced to glide over heavily residential areas before making a safe landing at Greenpoint. A month later, on Election Day, he performed loops, spirals, and "quick and fancy turns" over the Statue of Liberty.

That winter Domenjoz embarked on an extensive exhibition tour. On December 11, he flew for a Goshen, New York, hospital benefit. In February 1916 he was flying in Richmond, Virginia, before moving on to Washington, D.C., for an appearance in connection with the Pan-American Conference.

The tour carried him as far south as Havana, then back to New York for a return engagement at Sheepshead Bay. During the summer of 1916 Domenjoz and Baxter Adams, a Curtiss pilot, swung through the Midwest demonstrating such military skills as bomb dropping and mock dog-fighting.

Domenjoz had not completely severed his ties with Blériot. He returned to France in the winter of 1916 to test fly SPADs at Pau. By May 1917 he was back in the United States for another exhibition season. Following wartime service as a civilian flying instructor at Park Field in Memphis, Tennessee, Domenjoz made one final barnstorming tour in 1919, then placed his Blériot in storage on a Long Island farm and returned to France, where he remained for seventeen years.[1]

When Domenjoz returned to the United States in 1937, he had little hope of ever seeing his Blériot again. But while visiting George McLaughlin, then editor of *Aero Digest*, he learned that the machine had been sold to an aeronautical museum at Roosevelt Field, Mineola, Long Island, several years before to cover the unpaid storage costs. McLaughlin later recalled that "Domenjoz was practically in tears of joy as I told him this, and left immediately for the field to see 'his beautiful little airplane' as he called it."[2]

The "beautiful little airplane" remained in the collection at Roosevelt Field until February 1950, when Paul Edward Garber, then curator of the National Air Museum, was able to purchase the Blériot along with a Nieuport Scout that had been flown by Charles Nungesser; Thomas Scott Baldwin's *Red Devil*; and an assortment of aero engines for $2,500.

It was a particularly important acquisition. Only a handful of the hundreds of Type Onze aircraft produced by Blériot between 1909 and 1914 have been preserved. The Domenjoz machine was one of the last of its kind and had been received in very good condition with few alterations. It was a factory-original Blériot XI, representing the final fully developed configuration. For this reason a careful study of the restoration of the machine is of special interest to the student of pioneer aircraft. The decision to restore the Domenjoz Blériot was not made lightly. It was one of the first aircraft chosen for display in 1978 when planning began for a new "Early Flight" exhibition gallery at the Smithsonian Institution's National Air and Space Museum. At the time the Blériot was on exhibit in a no-frills display at NASM's Paul E. Garber Preservation, Restoration, and Storage Facility in Silver Hill, Maryland.

On the surface the machine appeared to be in sound condition. For that reason, and in the hope of preserving as much of the original craft as possible, including the fabric, NASM officials initially decided on a minimum restoration that would prevent further deterioration during the anticipated ten-year life of the exhibit. The fabric would not be removed from the wings, nor would the fuselage be disassembled.

Once work had begun, however, it became obvious that deterioration was more serious than had been supposed and a full-scale restoration effort was ordered. This included completely disassembling the aircraft and cleaning, repairing, restoring, or replacing parts as required.

As in all NASM restorations, the goal was not to return the machine to flight status, a task which would require the replacement of much of the original structure for reasons of safety. Rather, the intent was to halt all deterioration, restore the original detail, and preserve the craft as an example of prewar Blériot technology. The general rule in such a project is to preserve wherever possible, restore with original materials and techniques when necessary, and replace a part only as a last resort.

In the case of the Domenjoz Blériot, comparatively few completely new parts had to be fabricated. These included:

1. All of the reeds on both wings.
2. Cord wrapping on the lower longeron.
3. The top of the cowl.

4. Two turnbuckles above the throttle.
5. Two leather seatbelt straps.
6. Lower spreader bar in the fuselage waist.
7. The upper right and left mating sleeves for fuselage longerons.
8. The "B" logo on the front cowl.
9. New fabric on all surfaces.

Work on the Blériot began on March 20, 1978, and was completed on January 15, 1979. A total of 3,375 man hours were devoted to the task. Edward Chalkley, chief of the Preservation, Restoration, and Storage Division, and Walter Roderick, shop foreman, assigned craftsmen John Cusak, Joseph Fichera, Karl Heinzel, and Robert Padgett to the restoration of the Domenjoz Blériot.

In addition to returning the aircraft to factory-original condition, the restoration process offered an ideal opportunity to record the details of the construction of this final version of the Blériot XI. And it offers us, after having traversed the entire history of the Type Onze, an opportunity to make a detailed examination of a single specimen, from nose to tail.

PROPELLER

The Domenjoz Blériot is fitted with a laminated propeller, probably a Chauvière, although no company identification has survived. It is built up of six laminations and has a diameter of 250 cm and a pitch of 160 cm.

In spite of the efforts of the Wright brothers, the Polish engineer Stefan Drzeweicki, and others, propeller design remained an empirical business into the 1920s. Three to twelve hardwood laminations, usually mahogany, walnut, or ash, were glued and pinned together, each layer slightly offset from the one beneath it. The individual planks were often precut to the shape of a number of transverse sections, but all final shaping and finishing was done by hand. Even when copy lathes came into use in propeller manufacture prior to World War I, the finished product owed as much to the craftsman's artistry as to engineering science.

The propellers of the period had to stand a variety of stresses. As early as 1909 the Chauvière Intégrale propeller of the cross-Channel Blériot XI, which had a diameter of 2.08 m and produced 105 kg of thrust at 1,450 rpm, was already operating at a top speed of 185 m per second, close to 700 kph. At these speeds a propeller had to stand both enormous forward bending loads as a result of the thrust and heavy centrifugal loads created by rotation. The rough running engines of the period compounded the problem, creating frictional heat between the laminations and "burning out" a blade from the inside out.

Between 1909 and 1914 a great variety of propellers were flown on Blériot machines, but the fine blades produced by Lucien Chauvière, or the Normale propellers manufactured by M. Ratmanoff on the basis of Stephan Drzeweicki's patents, remained the most popular choices throughout the period.

ENGINE

To power his Blériot XI, John Domenjoz chose a Gnôme Type Sigma engine. Priced at $650, the Sigma developed 60 hp at 1,200 rpm. It had a bore and stroke of 120 mm and weighed 79 kg empty.

As noted in the text, the cross-Channel Blériot had originally been fitted with a 7-cylinder REP semiradial engine that was replaced by the famous

Dimensional drawings of a Blériot XI prepared for Charles Hayward's Building and Flying Aeroplanes *(Chicago, 1918).*

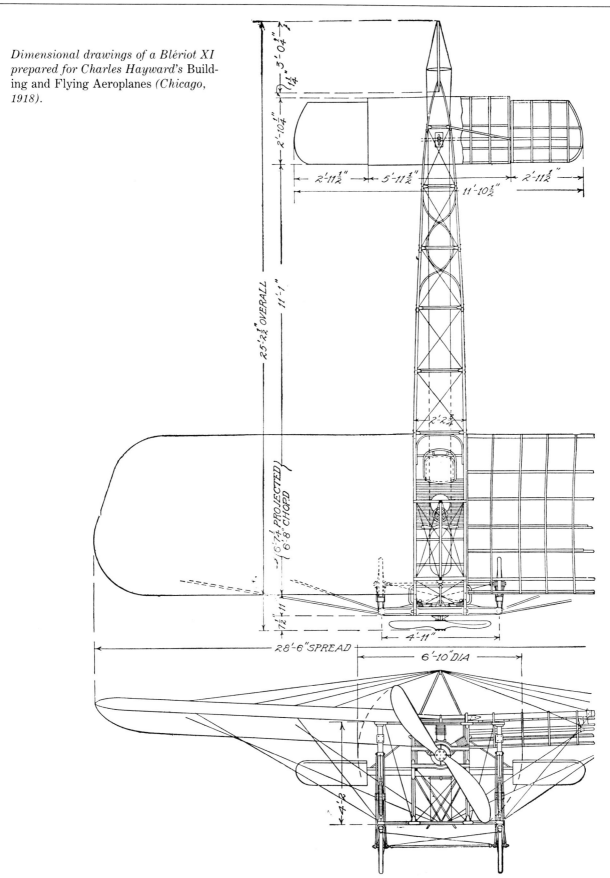

Fig. 23. Details of Early Model Bleriot Monoplane

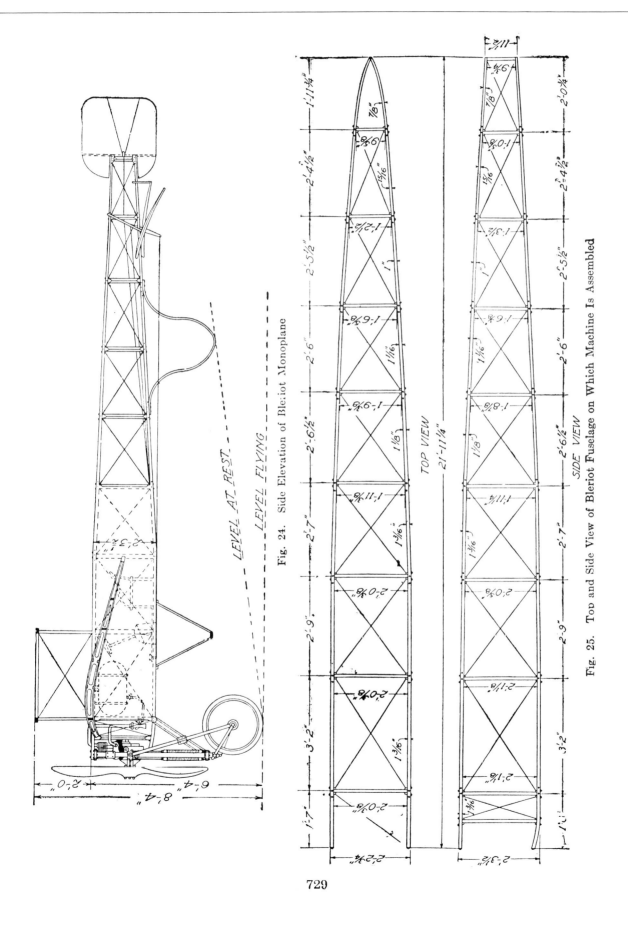

Fig. 24. Side Elevation of Bleriot Monoplane

LEVEL AT REST

LEVEL FLYING

TOP VIEW
21'-11¼"

SIDE VIEW

Fig. 25. Top and Side View of Bleriot Fuselage on Which Machine Is Assembled

25-hp Anzani powerplant in the spring of 1909. After 1910, Anzani-powered Blériots usually employed the uprated 35-hp version complete with magneto and overhead exhaust valves.

The appearance of the first Gnôme rotary engine in 1909 marked a major advance in aeronautical engine design. Produced by the brothers Louis and Laurent Seguin, the Gnôme was not the first rotary engine (a design in which the propeller and engine are bolted together and revolve around a fixed crankshaft). The Australian experimenter Lawrence Hargrave had first suggested such a powerplant in 1887 and the American machinist F. D. Farwell was operating a rotary as early as 1896. Stephen M. Balzer originally planned the powerplant of the 1903 Langley Aerodrome to operate as a rotary engine. But the Gnôme was the first genuinely practical, well-engineered rotary made generally available to aircraft builders.

As one contemporary noted: "If an expert engineer . . . were asked to examine a Gnôme motor . . . having no previous knowledge of the mechanism, he would unquestionably pronounce it an impractical, though highly ingenious construction."[3]

Indeed, the engine must have puzzled engineers encountering it for the first time. The gasoline, air, and the castor-oil lubricant were fed into the hollow crankshaft by a positive-acting piston pump. The mixture was then drawn into the cylinder from the crankcase through the inlet valve in the piston head. In order to counteract the centrifugal force of the rotating engine, these valves were fitted with counterweights.

The crankshaft turned on annular ball bearings at the front that helped absorb the propeller thrust. A second set of ball bearings was used on the master connecting rod. Plain bearings were employed on the auxiliary connecting rods and the piston connections.

After combustion the spent gases and liberal amounts of castor oil exited through exhaust valves on the cylinder heads. The exhaust valves were actuated by external rocker arms operated by cams. An "obdurator," a thin brass ring resembling a leather hydraulic seal, was substituted for conventional piston rings.

A gear-driven Bosch high-tension magneto located next to the oil pump at the rear of the crankcase provided power to a distributor with seven ebonite points, each linked to a spark plug in the cylinder head by an external wire. Under normal operating conditions, spark-plug fouling was not much of a problem, as the combination of the explosion and centrifugal force effectively cleared the cylinders. However, when the plane was descending with the ignition off, raw fuel and oil accumulated in the cylinder, and fouling did occur. For this reason, pilots were reluctant to cut the ignition for prolonged periods, preferring to blip the engine on and off during descent to clear the cylinders.

Gnôme engines were expensive, in terms of both the purchase price and the cost of constant upkeep. Earle Ovington kept three full-time French mechanics to service his Gnômes and insisted on a complete overhaul every 15–20 hours. "The valves were reground and retimed, new valve springs were inserted, the tappet rods were adjusted, and the whole motor was given a rigid inspection."[4]

Nevertheless, the Gnôme rotary offered a number of obvious advantages, the greatest of which was the high power-for-weight ratio. The radial cylinder arrangement made possible the use of a much shorter and lighter crankshaft than was required for inline engines of equal horsepower.

The additional weight of coolant, radiator, fans, water pumps, and the like could also be dispensed with. While fixed radial engines like the Anzani had also been air cooled, they had a much greater tendency to overheat than did the rotary Gnôme. Finally, the weight of the spinning engine acted as a flywheel that guaranteed smooth operation and prevented propeller damage.

Another reason for the popularity of the Gnôme was the obvious care taken during construction. To insure strength, very few castings were used. Almost every component was machined from solid drop-forged blocks.

The precision of Gnôme machining was legendary. Every spare ounce of metal was trimmed away, so that it was often remarked that the factory produced more steel shavings than engines. The piston walls were milled so thin that they could be crushed by hand. The rough crankcase casting weighed over one hundred pounds when it came from the forge, then was machined down to only 13½ pounds for the finished engine.

The most up-to-date German and American machine-tools and shop practices were employed to reduce production time and costs. Three hours were originally required to machine the original 80-pound solid cylinder block to the 4½-pound finished product. Through the use of new machines, this time was reduced to 7½ minutes, including the time required to cut the cooling fins only ½₀th of an inch thick, which were an integral part of the cylinder. The finished product was not only the most reliable aero engine on the market; it was a work of the machinist's art.

Gnômes were produced in both single- and double-cylinder row models ranging from 60 to 160 hp. Domenjoz's choice of the 60-hp Type Sigma is a bit odd. Considering the large wing area of his machine and the fact that he planned to earn his living as a stunt pilot, he might easily have chosen a more powerful engine.

As was the case with the rest of the aircraft, restoration of the Gnôme Sigma was a fairly straightforward task. The engine was well preserved with castor oil and showed little corrosion or rust. With the replacement of a valve spring and a few other small parts and a complete cleaning, the powerplant was ready for reassembly. Engine parts were coated with a mixture of CRC 3-36 and Soft Seal for lasting protection. Should the need ever arise to run the engine, these preservatives can be removed and normal lubricants added.

FUSELAGE

Like all Blériot Type XI aircraft, the fuselage of the Domenjoz machine consists of a wire-trussed box girder composed of four ash longerons united by thirty-five ash struts. Because of the difficulty of obtaining beams of the appropriate length, the longerons are made in two sections, joined in the middle by a 4.0 cm, 20-gauge steel ferrule, or sleeve, held in place by two ⅛-inch bolts. Each longeron varies in thickness from 2.0 cm square at the tail to 3.0 cm near the cockpit. The ash struts run from the single rear sternpost forward, dividing the fuselage into six bays or boxes behind the cockpit. Each strut is squared on both ends to mate with the longeron and is shaped into an ellipse in the center section. The sternpost measures 5.5 cm square on the ends. The remaining crossbraces run from 3.5 cm square on the end bay to 5.0 cm square behind the cockpit.

The struts and longerons are joined by a unique U-bolt fitting devised by Blériot to speed production. The smaller struts to the rear are held by ⅛-inch U-bolts; the larger forward braces use 3/16-inch bolts. All bolts measure 5.0 cm between the ends. Two bolts, one vertical, one horizontal, are used at each junction. Each strut is slotted with a 5/32- or 7/32-inch bit to receive

The fuselage during restoration.

Fuselage longeron, strut and bracewire arrangement. Note the twin U-bolt attachment system.

a U-bolt. The vertical struts are positioned one inch forward of the horizontal struts, and the two are then linked to one another and to the longeron with two U-bolts.

The fuselage is trussed diagonally with 20–25 gauge, 1.5–3.0-mm music wire. A turnbuckle connects each end of every wire to a protruding corner of a U-bolt. Original factory Blériots require roughly 100 turnbuckles of the twin screw-eye type. Internal diagonal trusses are also employed at the end of each bay.

The longerons fit into the sternpost in a tongue-and-groove arrangement. Twenty-gauge sheet-steel straps fastened to the longerons with ⅛-inch bolts complete the arrangement.

In the cockpit area, the strut-and-longeron arrangement is a bit different, since the cockpit must be free of struts and brace wires. A strip of 1.25 cm by 2.50 cm ash or hickory, bent into a U-shape, runs horizontally from one upper longeron (located behind the pilot's back), where it is bolted to the last upper horizontal rear strut, and around to be bolted to the other upper longeron. This brace was steam bent with a 23.0-cm radius curve. Two ⅛-inch bolts were used on each longeron and to fix the piece to the rear strut.

The cast-aluminum socket to hold the root ends of the rear wingspars is attached to the vertical struts just forward of the pilot's seat. The spars

The forward fuselage. Note engine and tank arrangement.

are held in place by a single bolt so that they are free to rock up and down for warping. The top horizontal strut at this point is a piece of steel channel, slightly arched over the control stick. A wooden strut is carried in the channel. This metal crossbrace is bolted through the rear spars as they pass through the aluminum casting on the vertical members. The top horizontal brace of the next bay forward is also steel and arched to pass over the top of the fuel and oil tanks. The instruments are mounted on this member.

A large, thick steel tube serves as both the top horizontal brace for the final bay before the engine compartment and the carry-through for the heavy forward wingspars. The engine itself is mounted between two steel plates forged into the shape of an X.

The top of the fuselage back to the cockpit is covered by a sheet-metal cowl to prevent the pilot from being splashed with excess oil. The cowl has three holes on top. The one on the right is the gas fill for the right tank. The other two on the left are for gas-and-oil fill for the left tank. Convex sheet-steel panels cover the sides of the engine and extend back toward the cockpit.

Blériot's fuselage-construction techniques remained standardized throughout the years of Type XI construction. Certain elements, notably the use of U-bolts to link struts, longerons, and brace wires, were patented.

Karl Heinzel works on the Blériot wing.

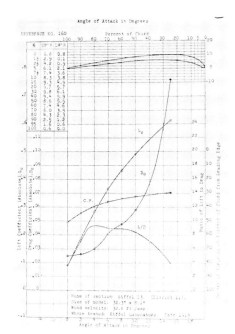

The Blériot XI (Eiffel 13) airfoil.

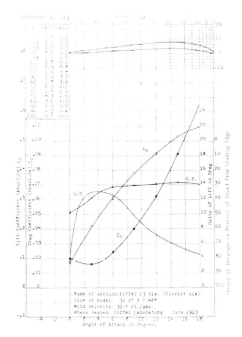

The Blériot XIa (Eiffel 13a) airfoil.

Detailed drawings of the Blériot wing, from L'Aérophile.

Restoring the fuselage of the Blériot presented few problems for the NASM craftsmen. The structure was completely disassembled and all metal fittings were removed for cleaning and treatment with preservatives. The longerons were wrapped with grade A cotton fabric and doped. Presumably the fabric wrapping, which is not present on all Blériot machines, was intended to strengthen the spars. The longerons and fuselage guywires were finished in white to match photographs of the Domenjoz machine.

WINGS

While the planform of Type Onze remained fairly constant throughout the period, internal wing construction was the result of a process of slow evolution. From the outset, the wing structure was the weak point of Blériot design.

The wings of the 1909 cross-Channel Type XI had been fairly simple. The span was 8.53 m with a chord of 1.83 m and a surface area of 13.95 square meters. The camber was roughly 1 in 20, and the aspect ratio was 4.65.

The classic Blériot XI airfoil was tested by the pioneer French aerodynamicist Gustav Eiffel and is identified as an Eiffel 13. Eiffel also tested a foil from an aircraft that he referred to as the Blériot XIa (the Eiffel 13-bis). The shallower camber suggests that the Eiffel 13-bis was used on racing aircraft. Standard test data for both airfoils are included in the accompanying graphs.

As noted, the rear spar was set into a cast-aluminum bracket in the cockpit area. The front spar fit into a metal carry-through. Twelve ribs spaced seven inches apart connected each pair of spars. The ribs varied in construction. The root ribs were built-up channels with separate web-and-cap and base strips. Moving outboard, other ribs were constructed of slender strips of wood, 6.5 mm square. Still others were simply cut from sheet aluminum with a wooden leading edge.

The wings were double surfaced, covered with rubberized Continental fabric. The flying wires were attached to each spar, guyed to a pyramid

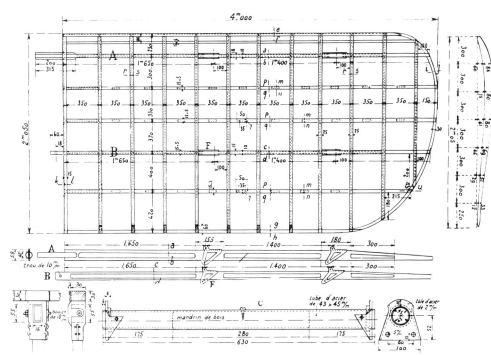

The Domenjoz Blériot after restoration.

constructed of steel tubing on the underside of the fuselage. The landing wires were guyed to twin steel-tube triangles located between the cockpit and engine.

The weakness of this original Blériot wing in compression and torsion has been described above. Over the next five years the engineers at Blériot Aéronautique gradually improved the design of the wings. The Domenjoz Blériot represents the successful result of their efforts.

The wings of the machine are Type XI-2 surfaces, apparently fitted to this single-seat machine to provide the additional lift required for stunt flying. Each wing has a span of 4.4 m and a chord of 2.0 m. The wing is built around two main spars of ash. The forward main spar measures 8.25 cm by 1.9 cm, while the rear beam is 6.35 cm by 1.5 cm. Both spars are planed to fit shaped slots in the ribs. Round stubs on the forward spar fit into the socket-tube carry-through, while the rear spar is still bolted to an aluminum-casting top of a fuselage strut.

In addition to the main spars, the leading edge constructed of thin plywood bent to shape adds additional stiffness. (In other Blériot craft, a four-inch strip of heavy sheet-aluminum, rolled around a piece of 5.75-inch half-round moulding, serves as a leading edge.) Three additional strips of spruce, 5.0 mm by 16.0 mm, and a spruce trailing edge, 6.5 mm by 3.2 mm, provide further stiffening.

Each wing contains thirteen ribs. Each rib is built of three pieces. The web is of 5.0-mm spruce with fine elliptical cutouts to save weight. Five additional holes are cut to accept the front and rear spars and the stiffeners. The top and bottom sides of the I-ribs are constructed of 5.0 mm by 16.0 mm spruce strips glued in place.

The ribs are spaced 35.0 cm on the centers and are tacked, rather than glued, to the beams. To assist in resisting drag, each wing is fitted with spruce strips running parallel to the ribs between the attachment point for the brace and warping wires. These strips are linked by two 3.0-mm cables, one running from the main spar near the root to the aft end of the first strut, the other from the forward tip of the first to the aft end of the second strip.

Each of the two spars on either side is furnished with sheet-steel fixtures for attaching the brace and warp wires.

The restoration of the wings was fairly straightforward. A number of cracked ribs were repaired by sandwiching the original wood between strips of 2.0-mm plywood. New leading edges were required on both wings. Additional plywood strengthening was required in other rib areas as well. All metal fittings were removed, glued, and finished with two coats of silver varnish.

The wings were recovered with grade A cotton fabric applied on the bias

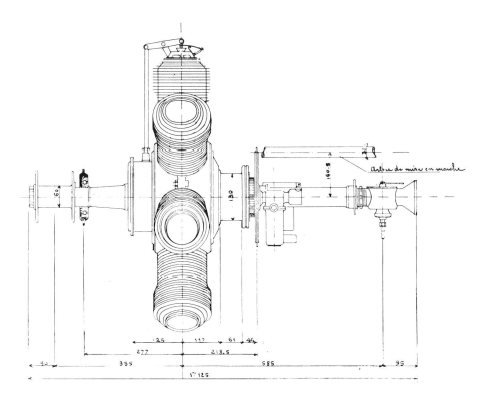

The Gnôme Sigma, side elevation.

to match the original. Blériot practice called for the fabric to be applied without stitching, using only glue and tacks. Thin wooden strips tacked over the ribs hold the cloth to the surface. The fabric was applied on the bias to avoid pocketing when the wing warping was operated.

The wings were finished with dope. Factory-original Blériot aircraft were usually given a coat of varnish as a final seal. This was omitted in the NASM restoration because of the fear that varnish might cause long-range damage to the fabric.

The tail numerals and the name DOMENJOZ were painted on the fabric with dope.

EMPENNAGE

The horizontal tail surfaces of the cross-Channel Blériot XI included a fixed center section with elevating tips. The tail was built around a steel tube bolted to the underside of the fuselage by cast-aluminum fittings. A flat steel strip bar provided major support for the tail, supplemented by a pair of light steel tubes.

A horn for the elevating tips was attached to the steel transverse tube and extended above and below the fabric at the center point. The tube, free to swivel in the center section, was fixed at the tips. The angle of incidence of the entire horizontal tail could be altered by means of a bracket on the trailing edge.

The rudder of the original Blériot XI pivoted some thirteen inches behind the trailing edge of the tail. It was a "high" rudder with an extension above the fuselage. Virtually all other Type Onze machines, including the Domenjoz Blériot, featured a square rudder with rounded corners and no extension. Most Type XI-2 aircraft had rudders mounted entirely on top of the fuselage as did a number of other Type XI derivatives.

The Domenjoz Blériot has a full-span elevator attached to a lifting tail of the sort that had first appeared on the Type XI cross-country machine of 1911. According to contemporary accounts this tail, which was to become standard for all Type XIs, was first developed by Paul I. Kuhling, a Blériot pilot. As in the earlier tail, the angle of incidence could be altered by means

Joseph Fichera and John Cusack fabric the wing.

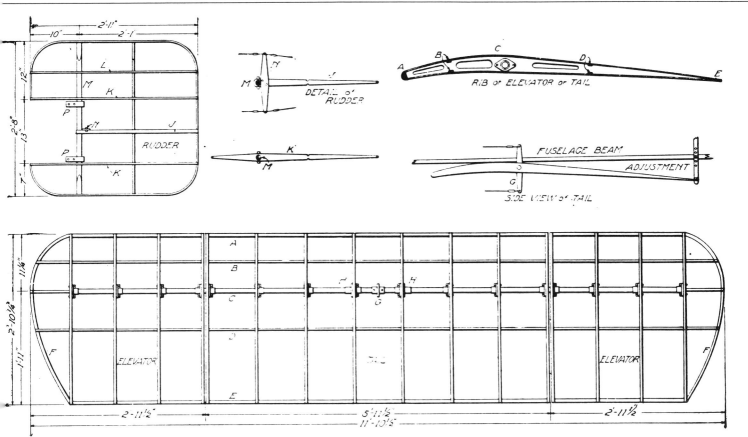

Empennage drawing, from Haywood.

Detail of restored undercarriage.

of a bracket at the center of the trailing edge. The elevator was operated by twin horns above and below the leading edge. It was braced by two steel tubes on either side.

The restoration of the empennage involved the same wood-, metal-, and fabric-working techniques employed on the wings and fuselage. No special problems were encountered.

UNDERCARRIAGE

The Blériot undercarriage was one of the most distinctive features of the aircraft and underwent few alterations during the course of production. The central feature was a bedstead frame on the nose of the machine which doubled as a forward engine and undercarriage support.

The bedstead consists of two horizontal beams, two vertical beams, and two vertical tubes.

The horizontal beams are ash, 12.0 cm wide at the center, tapering to 9.5 cm at the end and measuring 1.6 m in length. The upper beam is 5.0 cm thick, the lower 2.5 cm.

The vertical struts stand four feet two inches tall and measure 3.0 cm by 7.5 cm. Each end has a square tenon to mate with the horizontal beams. Ash struts run from the lower end of the uprights to the fuselage. Steel tape and cable are used to truss the bedstead.

The wheels are mounted in two forks. One of these serves as a radius rod while the other is attached to a slide held in place by steel-coil tension springs which absorb landing shocks. The wheels are free to swivel for cross-wind landings.

The rear wheel of the original Blériot XI gave way to a double U-shaped skid constructed of bent strips of hickory and ash some eight feet long. This

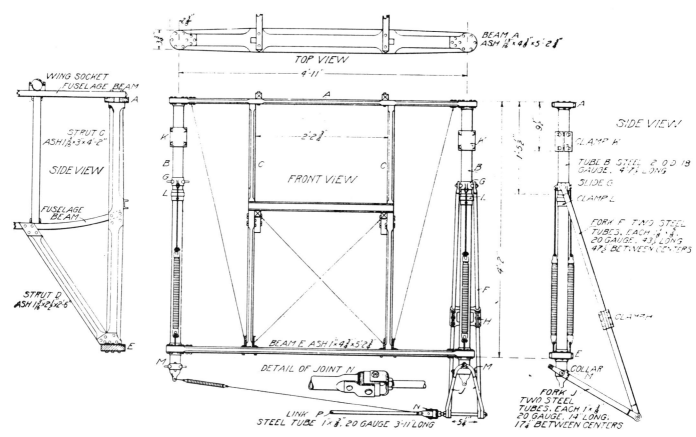

Undercarriage drawings from C. B. Haywood, Building and Flying an Aeroplane *(New York, 1911).*

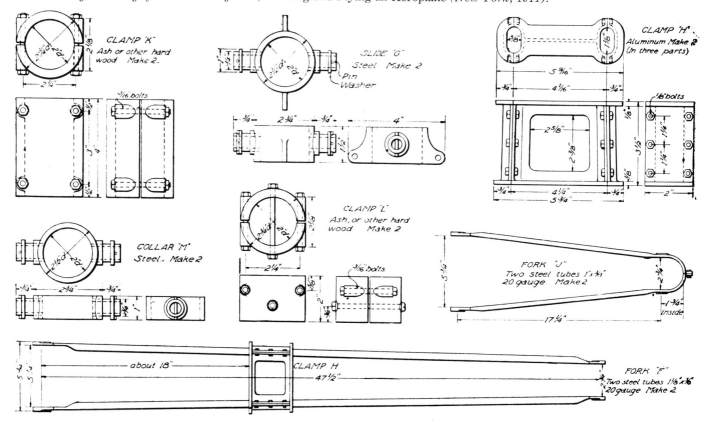

Undercarriage hardware, from Haywood.

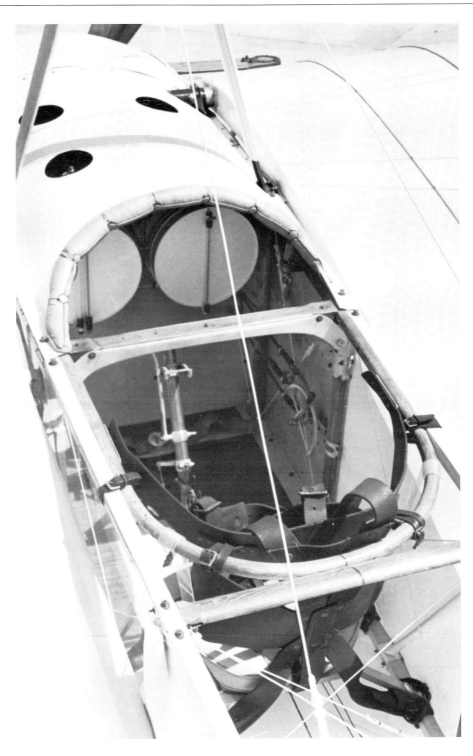

The cockpit area. Note tank sight-gauges and oil pulsator.

new skid, which first appeared in 1910, greatly reduced the long and dangerous landing runs of earlier machines.

The undercarriage was completely disassembled, cleaned, and treated with preservative during restoration. No replacement parts were required, but as with so many restorations of pioneer aircraft, tires presented a problem. As no tires could be found to fit the original wheels, a substitute set of near-original size was mounted on the aircraft. The search for a set of tires of the correct size continues.

The Domenjoz Blériot restored.

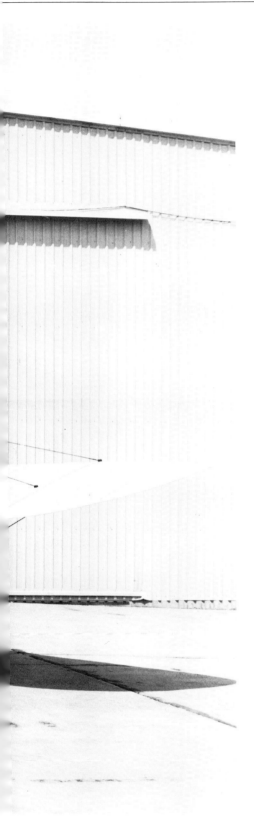

COCKPIT AND CONTROL SYSTEM

The pilot's seat in the Blériot XI consists of a cane bottom and light wooden side pieces pierced with holes to reduce the weight. The seat is supported by two steel bars bolted to the lower longerons. In the case of the Domenjoz craft, a special leather harness is provided for the pilot.

Standard Blériot controls are employed. The rudder pedal is a piece of shaped hardwood. Rudder cables (2.5 mm) doubled on each side like all the control cables are attached directly to the pedal and run back through the center of the fuselage to the attachment horns. Amateur builders often crossed the cable to provide more natural rudder control. In factory-original machines, the cables were not crossed, however, to make the use of the rudder for lateral control easier. The Blériot engineers were fully aware of the fact that when the machine was traveling at much under 40 mph, the wing warping was ineffective and the rudder became the primary means of obtaining lateral balance.

The control stick has a small wooden wheel some 20.0 cm in diameter mounted on top. The wheel is only a handle for the stick. It has no control function. A white blip button is mounted on the stick beneath the wheel.

In the original cross-Channel model the stick was attached to a cloche, a bell-shaped piece of metal to which two warping cables (3.2 mm) and two elevator cables (2.5 mm) were attached. A side-to-side movement dropped the appropriate wing, while a fore-and-aft movement operated the elevator. Later Blériots like the Domenjoz machine operated without the cloche. The stick operates the elevator directly. For warping, the stick is fixed to a geared wheel. The central section of the warping cable is a chain which meshes with the teeth of the gear.

The cockpit is well instrumented for the period. A sight gauge mounted on the back of the right-hand gasoline tank gives a reading on fuel. Twin sight gauges on the left tank indicate oil and gas remaining in that tank. A small handle on the pilot's left permits him to close both main tanks and draw fuel and oil from a small football-shaped tank mounted behind the engine. As the main tanks empty at the bottom and the small tank empties from the top, the pilot presumably closes the main tanks and draws from the small tank when flying inverted, a Domenjoz specialty. An air lever on a quadrant attached to the stick and a fuel lever and quadrant on the right wall have to be manipulated together to achieve maximum rpm. An oil pulsator above the fuel lever allows the pilot to see that oil is moving through the system. In addition, by counting the bubbles of oil in the glass of the pulsator, a pilot can estimate rpm.

Three instruments—a tachometer, a fuel-pressure gauge, and a pressure altimeter—are mounted on the fuselage crossbrace above the tanks.

New leather straps were produced for the bulky pilot harness designed to hold Domenjoz in place during aerobatic maneuvers. All other cockpit fixtures are original.

Today the finished product, embodying the combined craftsmanship of Blériot Aéronautique workmen and NASM's expert restoration specialists, is the centerpiece of "Early Flight," a major museum gallery. It is a beautifully preserved example of pioneer aircraft technology, a fitting tribute to the classic Type Onze and the gallant aviators who flew it.

A Blériot Chronology

The period prior to the outbreak of World War I was a time of such rapid aeronautical advance that few designs survived through the entire era. Yet the Blériot XI, as a result of structural improvements, powerplant changes, and other alterations, remained in active use from 1909 through 1914.

During the early years, from 1909 to 1912, the Type Onze dominated aero meets and races and set records for altitude, speed, distance, and duration. The year 1912 was critical, during which the basic airworthiness of the machine was called into question. It was followed by the Blériot's re-emergence as the world's first aerobatic craft. Finally, the Blériot XI, which had formed the foundation of many early European air forces, made the first reconnaissance flights in the Balkan and Libyan campaigns and was in active service with the French and English air services during the first year of the war.

Rather than attempt to present a detailed narrative account of the achievements of the Blériot XI and the people who flew it, the following chronology will provide a capsule view of the development, accomplishments, and influence of the design.

December 24, 1908
    The Blériot XI is unveiled at the First Paris Aeronautical Salon.

January 23, 1909
    The first flight of the Blériot XI, 200 meters, at Issy-les-Moulineaux.

March 9–15, 1909
    Continued flights at Issy, including the first turns.

May 27, 1909
    First flight with the 35-hp, 3-cylinder radial engine designed by Alessandro Anzani.

June 1909
    Hubert Latham announces his intention to fly the English Channel.

June 1909
    Louis Paulhan flies the first Gnôme engine, constructed by Louis and Laurent Séguin, on his Voisin.

July 1909
    Blériot flies the XI and XII at Douai and Juvisy.

July 2, 1909
    Blériot and Voisin are awarded the Osiris Prize.

July 3, 1909
    Blériot remains in the air for 48 minutes, covering 26 miles in the XII. His foot is badly burned by the engine exhaust.

July 13, 1909
    Blériot flies from Etampes to Cheville and back, 26 miles in 45 minutes, to win a 14,000-franc prize for a flight of over 40 km.

July 19, 1909
    Hubert Latham attempts his first flight over the Channel.

103

Blériot in a Type XI on the first day of the Reims Meet, August 1909.

July 21, 1909
Blériot arrives in Calais and begins preparations for the Channel crossing.

July 25, 1909
Louis Blériot flies the English Channel and wins the £1,000 *Daily Mail* prize.

August 24, 1909
Louis Blériot sets a world speed record of 46.18 mph. Four days later he raises it to 47.84 mph.

August 28, 1909
Blériot is awarded the Tour de Piste at Reims.

September 1909
Blériot now has 100 orders for the Type XI.

October 17, 1909
Blériot makes the first flight in Hungary, at Budapest.

October 18, 1909
In a driving rain, Hubert Leblon wins the Bradford Cup and the Doncaster Meet for a flight of 22 miles in 30 minutes.

October 26, 1909
Delagrange reaches a speed of 49.9 mph for a world record at the Doncaster Meet.

October 30, 1909
Blériot makes the first flight in Rumania, at Bucharest.

November 1909
Rodman Wanamaker brings a Blériot XI to the United States.

December 10, 1909
Blériot suffers a near-fatal crash at Istanbul.

December 11, 1909
Jacques Balsan wins the Prix Robert Demoulin in a Blériot XI.

December 30, 1909
Delegrange flies 124 miles in 2 hours 32 minutes, averaging 48.9 mph.

1909
Flying at Pau, Miss Edith Cook becomes the first Englishwoman to solo. Baron Carl Cederstrom is the first Swede to earn a brevet, also at Pau.

A typical Blériot race scene.

January 24, 1910

Delagrange dies in a crash at Bordeaux—the first Blériot fatality.

February 1910

Balsan wins first prize in a 5-km race at Heliopolis. Leblon, also flying a Blériot, wins the 10-km event.

February 10, 1910

Julien Mamet gives the citizens of Barcelona a firsthand look at a Blériot.

March 10, 1910

Emile Aubrun makes one of the world's first night flights, 12.4 miles, at Buenos Aires.

April 2, 1910

Hubert Leblon dies in a Blériot crash at San Sebastian, Spain.

Molon flying a Blériot at Lyon.

Jan Olieslagers's Blériot.

April 27, 1910
Mamet makes the first flight in Portugal, at Belem.

April 30–May 5, 1910
Léon Molon flies in the Grande Semaine de Tours.

May 6, 1910
Jan Olieslagers, a Belgian airman, flies at Barcelona.

May 8–16, 1910
Léon Morane wins the altitude prize at a St. Petersburg, Russia, air meet.

May 15, 1910
Olieslagers flies at Genoa.

May 15–17, 1910
Molon participates in a meet at Lyons.

May 21, 1910
Jacques de Lesseps flies from Calais to Dover in 37 minutes to win the £500 Ruinart Prize and a £100 *Daily Mail* Cup for being the second man to fly the Channel.

May 27, 1910
Bartolomeo Cattaneo wins the Prix de Vitesse at the Verona meet.

June 1910
James Radley makes the first public flights in Scotland.

De Lesseps crossing the Channel to win the Prix Ruinart.

June 1910
  Cattaneo wins the speed and duration prizes at a Rouen meet.

June 1910
  Morane sets a world monoplane record for duration flight with a passenger, 85 km.

June 3–6, 1910
  Aubrun and Balsan fly at an Anjou meet.

June 5–12, 1910
  Elie A. Mollien and René Barrier fly in the meet at Mondorf-les-Bains.

June 9–16, 1910
  Ferdinand Deletang participates in the Grande Semaine de Paris at Port-Aviation.

June 19–26, 1910
  Morane wins the altitude prize at the Rouen meet.

June 27, 1910
  De Lesseps makes a 30-mile flight around Montreal at an altitude of 2,000 feet.

July 1910
  Blériot pilots take the lion's share of the prizes at the second Reims meet. Leblanc wins the elimination race for the Gordon Bennett Tro-

Leblanc's "Circuit de l'Est" Blériot.

phy, as well as the M. Ephrissi prize. Aubrun wins the passenger-carrying prize, flying 85 miles in 2 hours 9 minutes, while Mamet takes second place for a flight of 57 miles with 2 passengers. Léon Morane is awarded the speed prize for a world record flight of 66.18 mph. Olieslagers wins awards for total distance flown and for a duration flight of 243.5 miles in 5 hours 3 minutes.

July 10–16, 1910
Morane dominates the Bournemouth meet, winning honors for general merit, altitude (4,478 feet) and speed. J. A. Drexel also places well at Bournemouth.

July 23–August 4, 1910
At a meeting in Brussels, Olieslagers wins the altitude (5,286 feet) and duration prizes. M. Tyck also flies a Blériot at the meet.

July 26, 1910
De Lesseps flies 40 miles from the Ile de Gros Bois in the St. Lawrence River.

August 1910
It is announced that five of the nine pilots who have flown to an altitude of over 1,000 meters (Morane, Chavez, Tych, Olieslagers, and Drexel) have done so in Blériot aircraft.

August 1910
Leblanc and Aubrun take first and second place in the Circuit de l'Est race sponsored by *Le Matin*. This is the first of the great long-distance air races.

August 1910
Morane captures first place for cross country, speed, and altitude at the Caen Meeting. Aubrun is also a prizewinner.

August 6–13, 1910
J.A. Drexel sets a new altitude mark of 6,600 feet at Lanark. Cattaneo wins the award for total distance flown, while Radley captures the speed prize. James McArdle and his Blériot are also award winners.

August 10, 1910

Claude Grahame-White carries the first "unofficial" English airmail from Blackpool.

August 14–21, 1910

M. Taddeoli, E. Dufour, M. Mouthier, and H. Amerigo fly Blériots at a Geneva Meet.

August 14–21, 1910

Morane and René Simon dominate the speed and altitude events at the Nantes meeting.

August 16, 1910

John Moisant completes the first two-man flight across the Channel with passenger Albert Fileux in a Blériot XI-2.

August 24, 1910

Cederstrom flies 30 km over Copenhagen Sound, from Amagar, Denmark, to Limhamnsfeltet, Sweden.

August 26–September 6, 1910

Blériot wins the constructors' award at the Bois de Seine meet. Morane captures the altitude and speed prizes, and Simon wins first prize for the greatest distance flown.

September 1910

Léon Morane wins the altitude, speed, and cross-country awards at Bordeaux. Aubrun wins the 1910 Coupe Michelin for a nonstop flight of 196 miles.

September 3, 1910

Morane sets a world altitude record of 8,471 feet at Deauville.

September 3–13, 1910

Claude Grahame-White wins the $1,000 Boston Globe award for a flight around the Boston Light in 33 minutes, 1⅕ seconds.

September 8, 1910

Georges Chavez sets a new world altitude record of 8,488 feet.

September 11–18, 1910

Emile Aubrun, Simon, Morane, Mollien, G. Legagneaux, Tyck, and R. Gilbert fly their Blériots at a Bordeaux meet.

September 22–25, 1910

Comte de C. Boise and M. Mauthier fly at a Dijon meet.

September 23, 1910

Chavez dies in a crash following the first flight over the Alps.

September 27–30, 1910

Morane is the star of a flying meet at San Sebastian, Spain. He explains the details of his machine to King Alphonso.

October 1910

Drexel sets a new world altitude mark of 9,449 feet at Philadelphia.

October 1910

Cattaneo wins the largest share of the prize money at a Milan meet. Legagneaux wins the altitude prize. Marcel Paillete, Simon, Tyck, Aubrun, and Wienczers also fly Blériots.

October 5, 1910

Léon and Robert Morane are injured when the wing of their Blériot collapses at Boissy St. Leger.

October 12, 1910

Leblanc sets an unofficial U.S. record of 78 mph at St. Louis.

October 22–31, 1910

De Lesseps wins the $10,000 prize offered by Allen A. Ryan at New York's Belmont meet. Claude Grahame-White wins the $5,000 Gordon Bennett Race on October 29, covering 62.1 miles in 63 minutes, 4⅗ seconds.

October 29, 1910
Leblanc sets a new world speed record of 79 mph at Belmont.

October 30, 1910
Grahame-White and Moisant battle for a $10,000 prize in a race to the Statue of Liberty.

October 31, 1910
Drexel reaches 8,370 feet at Belmont.

November 1910
Japanese military officials visit Issy for a flight on the two-seat Blériot recently purchased by their government.

November 1910
Yves Guyot demonstrates his Blériot in St. Petersburg and Moscow.

November 1910
A Blériot is reported to have been sent to Persia to assist in surveying operations.

November 23, 1910
Drexel climbs to 9,714 feet over Philadelphia.

December 1910
Louis Blériot takes his wife, Alice, for a flight in the new two-seat Type XIV.

December 17, 1910
Cattaneo wins $20,000 for a 70-mile flight across the River Plate from Argentina to Uruguay and back.

March 1911
Earle Ovington imports a new Blériot to the United States. It is quickly copied by local manufacturers.

April 12, 1911
Pierre Prier becomes the first man to fly from London to Paris (3 hours, 56 minutes). Prier is chief flying instructor at the Blériot school at Hendon.

May 28–June 1, 1911
Jean Conneau wins the Paris-to-Rome race. Roland Garros, also in a Blériot, finishes second.

June 12, 1911
Leblanc flies at 77.67 mph, a new world record, at Etampes.

June 18–July 7, 1911
Conneau wins the $32,300 Circuit of Europe prize.

July 22, 1911
Conneau wins the £10,000 *Daily Mail* Round Britain race. Once again, Garros finishes second in his Blériot.

July 25, 1911
Alexandre de Wassilief wins the 1,000-km race from St. Petersburg to Moscow with a flight time of 25 hours.

August 1911
Three Blériots flown by Italian officers Piazza, Ginocchio, and Roberti form the core of an Italian operational air group in Libya. A few Blériots remain in operational service there as late as 1922.

September 1911
Andriani sets new Russian records for altitude and duration.

September 4, 1911
Roland Garros reaches 13,943 feet near St. Malo.

September 9, 1911
Gustav Hamel carries the first official English airmail from Hendon to Windsor.

September 14, 1911
Earle Ovington, operating out of a temporary post office at Sheepshead Meadow, New York, carries the first U.S. airmail in his Queen Blériot.

September 17–21, 1911
Leblanc reaches an unofficial speed of 81 mph in his Blériot XXVII.

September 29–October 3, 1911
Manissero and Verona place first and second in the Circuit de Milan. Another Blériot finishes in third place.

October 1911
Le Lasseur de Ranzay and Baron Della Noce fly 100 km from Bologna to Florence over the Apennines.

October 1911
Jean Raoult makes the first flight in Madagascar.

October 25, 1911
Carlo Piazza makes the first military reconnaissance flight over Turkish positions from Tripoli to Azizia.

October 31, 1911
Chave wins the Prix des Escales at Orléans. Dancourt finishes third in a Viale-powered Blériot. Lusetti also participates with a Blériot but does not place.

1912
STI (Società Tranaerea Italiana) in Turin begins to manufacture the Blériot XI.

April 21, 1912
Harriet Quimby becomes the first woman to fly the English Channel. Quimby had earlier become the first licensed woman pilot in America (August 1911).

June 8, 1912
Gustav Hamel and Trewhawke Davis win the Circuit of London race sponsored by the *Daily Mail*.

June 16–17, 1912
Roland Garros, back from his tour of Latin America with the Moisant International Aviators, flies to victory in the Circuit of Anjou, the last major victory for a Blériot.

July 1, 1912
Harriet Quimby and William Willard die in the crash of a Blériot XI-2 at Boston.

August 18–19, 1912
Edmond Audemars becomes the first man to fly from Paris to Berlin, via Bochum.

1912–1913
During the Balkan War, Bulgaria, Rumania, and Turkey all operate Blériot machines.

January 24, 1913
Oscar Bider makes the first flight over the western Pyrenees from Pau to Madrid.

March 1913
Lt. Alexandre de Kouzminski makes the first flight in Macao. He later makes the first flights in Siam and Cambodia as well.

Henri Salmet leaving Issy on the second leg of his London-Paris-London flight, March 1912.

March 11, 1913
Edmond Perreyon sets a new world altitude mark (19,291 feet).

June 3, 1913
Perreyon sets a new altitude mark for flight with a passenger.

June 15–22, 1913
Perreyon participates in the Semaine de Vienna meet.

August 19, 1913
Adolphe Pégoud tests the Bonnet parachute at Buc.

September 1913
Pégoud performs his first aerobatic maneuvers at Buc.

November 10, 1913–January 1, 1914
Jules Védrines flies from Nancy, France, to Sofia, to Istanbul, to Cairo, and back, a distance of 2,500 miles.

April 17, 1914
Hamel flies 340 miles from Dover to Cologne, crossing the Channel and five national borders in an XI-2 with a passenger.

July 30, 1914
Norwegian Trygve Gran makes the first air crossing of the North Sea in his Blériot.

August 12, 1914
Lt. R. B. Skene and R. K. Barlow become the first English airmen to die on active service during wartime when their Blériot XI-2 crashes.

August 19, 1914
Capt. P. Joubert de la Ferte completes the first RFC reconnaissance flight of the war. He is accompanied by G. W. Mapplebeck in a BE2b.

Flying the Blériot XI

The flying characteristics of the Blériot XI are likely to come as an unpleasant surprise to any pilot accustomed to the control response and precision of a modern airplane. Frank Tallman, who had some 120 hours in two replica Blériots powered by modern Aeronca and Continental engines, found it "hard to put into words readily explainable to any modern pilot how perfectly awful it is to fly the Blériot and what a great admiration I have for the pilots whose raw courage often far outstripped their piloting skills and knowledge."[1]

The problems begin on the ground, where the Blériot is just as tail-heavy as it is in the air. A minimum of two men, preferably three, is required to push the craft into position.

Preflighting the airplane is a fairly simple procedure, as Earle Ovington explained:

> First I looked at my big Gnôme motor. I turned it over slowly and felt the play of each exhaust valve. I examined carefully the bolts which held the main supporting steel strips to the landing chassis and the wings, in order to see that these fastenings were perfectly secure. The tension of these strips is also important and should be practically uniform. Very often a distortion of wings makes one of these strips tight and the other loose. . . . Don't forget to glance at the upper supporting wires, as often a turnbuckle may become loosened. . . .

> The leading edge of the wing should be absolutely parallel with the trailing edge. Squat down behind each wing and glance along these edges, and if they are not parallel, adjust the upper or lower wires until they are. This is of utmost importance, for the machine will not be on an even keel when the control is at a central position unless the two wings are adjusted perfectly equal. Don't forget to take a look at the tail and see that the supports holding it are firm, and the nuts and bolts secure.[2]

Care also had to be taken to check fuselage alignment, the condition of the bedstead and undercarriage, and the control wires. As one modern Blériot pilot warned: "The Blériot is a wire airplane and without each wire properly attached and safetied, it has about the strength of a fifteen-cent grocery-store kite."[3]

With the preflight complete, the pilot was ready for takeoff.

Engine starting procedures varied over the course of Type Onze production. In the case of the original cross-Channel machine, the "throttle" was a spark advance/retard lever on the right side of the stick. There were no

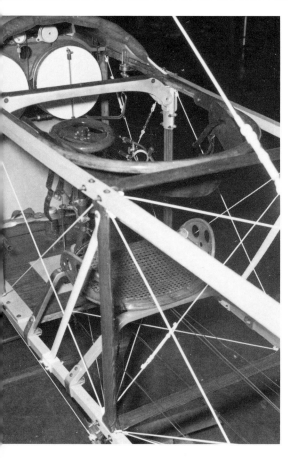

The cockpit of the Domenjoz Blériot XI.

instruments, except a sight glass on the oil line. The pilot counted the oil pulses as a rough gauge of rpm. An exhaust valve-lifter was located forward on the right-hand cockpit wall to stop the engine in flight. A bulb-operated oil pump on the right cockpit wall had to be operated by hand to maintain oil pressure.

To start, the carburetor of the Anzani was first flooded. Batteries were then attached to the sparkplugs and the spark lever was retarded. The propeller was swung through to draw gasoline and oil into the cylinders, the spark was advanced, and the propeller spun until the engine caught.

Starting the Gnôme engine of the 1914 Domenjoz Blériot was a quite different matter. A mechanic would inject gasoline and a bit of oil for compression into the exhaust valves of three or four cylinders. The propeller and engine were then spun through to start. The ignition system was always open, the engine always full on. The only engine speed control was the blip switch on the stick, which could be used to cut the ignition for some control of landing speed.

Once the engine was running, the pilot manipulated the air level on the control stick and fuel lever on the cockpit wall to achieve a correct mixture for maximum rpm. In the case of the Domenjoz Blériot, a tachometer was provided on the instrument panel, but the oil pulsator on the right cockpit wall could also be used to judge rpm. The pilot simply counted the number of oil squirts into the glass bulb per minute and referred to a table to get the rpm.

The novice pilot was now ready to taxi, not an easy task given the relatively narrow undercarriage, small rudder, and lack of keel area. Many a first flight ended in a ground loop, particularly with an Anzani engine which put little pressure on either the rudder or horizontal stabilizer.

Still, given the relatively effective Blériot airfoil, the airplane lifted quickly enough. With a 70-hp Gnôme, the tail was up within 50 feet and the machine off the ground within 200.

The aircraft was most vulnerable at takeoff and landing, when the air speed was between 20 and 30 mph. At these low speeds, there was virtually no lateral control. The pilot had only the rudder to maintain balance in roll during these crucial moments. Of course, at these low speeds even the rudder was only marginally effective, and on a gusty day the situation became virtually impossible. The relatively heavy stick forces that resulted from the stiff Blériot wing structure (particularly in later models) complicated the problem for the inexperienced pilot, who could easily be deluded into believing that he was feeling aerodynamic pressure when attempting to warp the wings.

Even under ideal conditions, with the aircraft flying at 40–50 mph, lateral control was marginal in gusty winds. Frank Tallman notes that "there is still no experience in my years of flying to equal the sick feeling you have when a wing goes down in gusty air and you head for the ground, unable to pick up the wing in spite of full opposite control."

Earle Ovington was careful to warn prospective Blériot pilots of the danger of side-slipping. "The trick is—do not let it get too far over," he cautioned, "but apply your corrective the minute the machine tilts to an undesirable angle."[4]

This feeling was shared by many monoplane pilots of the 1910–14 period. Jules Védrines, one of the most daring aviators of the era, was once accused of being afraid to attempt steeply banked turns. His response was to go up immediately and make the tightest turn he could around a pylon. When presented with a photograph of the event some time later, he apparently had second thoughts, for across the bottom of the picture he scrawled "Flight of an Imbecile."[5]

A number of pilots have also commented on the fact that the standard Blériot XI was tail-heavy, requiring a healthy forward pressure on the stick to maintain level flight. In the case of the two-plane Type XI-2, the condition was even more dangerous. Designed for flight with a passenger in the rear seat, the aircraft featured a lifting-tail set at a higher angle of attack than the wings. Pilots like B. C. Hucks, Gustav Hamel, and Harriet Quimby had a difficult time making the transition to this nose-heavy aircraft.

At least one observer blamed Harriet Quimby's fatal crash on the pitch instability of the XI-2. Quimby, an experienced aviator, had been the first American woman to earn a pilot's license and was the first woman to fly the English Channel in an airplane.

Just before 6:00 P.M. on July 1, 1912, she had taken off in her two-seat Type XI-2 Tandem, with William A. P. Willard, father of aviator Charles Willard, as passenger. The pair flew some eight miles from Squantum, Massachusetts, to Boston Light and were on their way back across Dorchester Bay, descending from 1,500 feet and banking to the left when the machine suddenly nosed down. Willard's body was thrown completely over the nose of the airplane. Harriet Quimby, "America's darling," was tossed out of the machine seconds later. The bodies and the wreckage of the Blériot fell into four feet of water one thousand feet from shore.

As usual, opinions varied as to the cause of the tragedy. Earle Ovington, who inspected the wreckage, informed reporters that the controls were jammed. But such evidence from a machine that had fallen 1,000 feet was scarcely convincing.

A more likely explanation came from the mechanic who had serviced the machine. "Too steep a glide," he remarked. "The machine lost its balance."[6]

Quimby had complained of the airplane's tendency to nose over. When flying solo, she had found it necessary to weight the rear seat with sandbags. Even so, up elevator had to be applied to keep the aircraft on an even keel.

The potential for disaster was obvious. If a passenger, particularly an overweight passenger like Willard, were to have leaned forward to shout in the pilot's ear, the sequence of events leading to the Squantum tragedy would have followed naturally. The aeroplane, already descending, would have nosed down suddenly with the shifting weight. Willard, leaning forward with no seatbelt, would have been tossed over the low fuselage as the Blériot immediately entered a dive. Quimby, shocked to see Willard go overboard and caught off balance as the nose plunged down, would have been thrown out of her seat as well.

Ross Browne, an American Blériot pilot, remembered that "landing was the hardest part of all." The elevator of the Blériot XI was the only control that was oversensitive. "Working your wheel back and forth even an inch would make a big difference," Browne recalled, "and it would be very easy to turn on your nose."[7] The machine had such a steep angle of descent (perhaps thirty degrees) that the stick could not be pushed forward and held; it had to be returned to neutral once the descent had begun. Because of the rapid drop, most pilots were reluctant to descend with the engine off.

Obviously, most of the rather frightening control problems cited above were not limited to the Blériot XI but were endemic to pioneer aircraft. These machines were built by men who were exploring a totally new technology. Their analytical tools were often weak, with intuition and common sense taking precedence in the absence of engineering principles that had yet to be developed. If the aircraft that emerged from this era could not yet offer the pilot complete mastery of the new environment, it could, at least, enable him to get into the air. And, for the time, that was quite enough!

Perhaps Air Commodore Allen W. Wheeler put the flying characteristics of pioneer aircraft in the best perspective:

> For any pilot accustomed to modern light aeroplanes, let alone high-performance military types, the most noticeable thing about flying 1910 aeroplanes would be the very considerable feeling of insecurity . . . After a little flying on the types it is possible to appreciate that the feeling of insecurity is mainly due to the general flexibility of the whole structure . . . and . . . that response to movement of the lateral controls is sluggish, sometimes to the point of nonexistence.

> Having mastered one's fears on all these points, it is possible to get joy, and certainly a thrill, out of flying 1910 aeroplanes. One is out in the open with no windshield in front, in fact very little all round; the whole flight is essentially a friendly arrangement between the pilot and the aeroplane: the former cannot *tell* the latter what to do; he can only arrange a situation wherein the latter will probably do what the former wants.[8]

Select Blériot Aircraft, 1901-1914

From the beginning of his work in aeronautics to the outbreak of war in August 1914, Louis Blériot produced, by his own count, some forty-five distinct aircraft designs. No complete record of the number of each type manufactured has survived, but it is apparent that, with the exception of the Type XI series, most were either strictly experimental machines or were produced in very small numbers.

In the absence of factory records it has not been possible to identify each of the forty-five designs. What follows is a checklist, based on existing secondary studies as well as contemporary journal and news accounts. The specifications given for each craft have been gathered from a wide variety of published sources and have been cross-checked to the extent possible. Nevertheless, the reader should recognize that there are discrepancies among sources.

In preparing such a list, one can only agree with the editor of the English journal *Aero*, who offered the following comment in 1911:

> In passing, it may be remarked that the system of distinguishing [Blériot] types by christening them with a numeral is apt to prove embarrassing, especially where, as in the present case, there are no known machines intermediary between XII and XXXI. No. XI we know; No. XI-bis we recognise; even Nos. XII and XIII we have heard of by repute. But where are XIV, XV, and their successors? Let us call No. XXI the "military two seater" and so rest in peace.[1]

In the list that follows, the Blériot model number precedes the year of construction.

I. 1901
An ornithopter powered by a lightweight carbon-dioxide gas engine. As many as three may have been built. The author is aware of only one illustration of the model, and it is not reproducible. The two wings are driven by a piston apparently fed from an upper reservoir. Bracewires are provided for the central section. Kingposts and guywires brace the wings. No details as to size or weight are available.

II. 1905
Constructed by G. Voisin based on his earlier Archdeacon glider. Biplane box-kite cells with side curtains fore and aft; forward elevator; mounted on twin floats. The upper wing of the forward cell is longer than the lower, so that the side curtains are angled in. See text for further details.
Span: 7.0 m (forward cell)
Area: 29.0 m² (forward cell)
Gap: 1.5 m (forward cell)
Weight: Less than 200 kg

III. 1905
Designed by Blériot and Voisin; built by Voisin. Elliptical wing cells fore and aft; mounted on floats. This was Blériot's first full-scale powered machine. See text for further details.
Area: 60.0 m²
Engine: 24-hp Antoinette
Propellers: Twin tractors, 2.0 m diameter

IV. 1905
Designed by Blériot and Voisin. A modification of both the Blériot II and III, with an elliptical canard wing and side curtains on the inside bay. Canard biplane elevator; mounted on floats. See text for further details.
Span: 10.5 m (forward)
Chord: 2.25 m
Gap: 1.85 m
Area: 47.0 m² (forward)
       26.0 m² (rear)
       78.5 m² (total)
Elevator span: 3.4 m
Elevator chord: .8 m
Elevator area: 5.5 m²

Weight: 430 kg
Engine: 2 24-hp Antoinettes
Propellers: 2.0 m diameter

V. 1907

Swept-wing canard monoplane with cockpit at rear; mounted on a tricycle undercarriage. No. V was the first Blériot machine to leave the ground for short hops of 2–5 m.
Span: 7.8 m
Chord: 2.0 m
Area: 13.0 m²
Length: 8.5 m
Weight: 260 kg
Engine: 24-hp Antoinette
Propeller: 1.6 m diameter

VI. 1907

A tandem-wing monoplane designed for Blériot by Louis Peyret, inspired by the Langley Aerodrome of 1903. Some long hops were made by both Peyret and Blériot. A 50-hp Antoinette was later installed. See text for details.
Span: 5.85 m (each wing)
Chord: 1.5 m
Area: 18.0 m²
Length: 6.0 m
Weight: 280 kg
Engine: 24-hp Antoinette

VII. 1907

Blériot's first step toward the classic tractor monoplane.
Span: 11.0 m
Chord: 2.5 m
Area: 25.0 m²
Length: 9.0 m
Weight: 425 kg
Engine: 50-hp Antoinette
Propeller: 4 blades, 2.2 m diameter

VIII. 1908

Blériot's first real success, and the direct precursor of the Type Onze. Details given below are for the original craft. See the text for details of the modifications to the VII-bis and -ter.
Span: 11.2 m (reduced to 8.5 m in VIII-bis and -ter)
Chord: 2.0 m
Area: 22.0 m²
Length: 10.0 m
Weight: 480 kg
Engine: 50-hp Antoinette
Propeller: 4 blades, 2.2 m diameter

IX. 1908

A monoplane exhibited at the Grand Palais in December 1908. The aircraft was tested but apparently did not fly.
Span: 9.0 m
Chord: 2.0 m
Area: 26.0 m²
Length: 12.0 m
Weight: 558.7 kg
Engine: 50-hp Antoinette
Propeller: 2.1 m diameter

X. 1908

A large biplane derived from the basic Wright configuration. Exhibited at the Grand Palais in December 1908. It was never flown.
Span: 13.0 m
Chord: 2.0 m
Area: 68.0 m²
Length: 8.2 m
Weight: 620 kg
Propeller: 3.0 m diameter

XI. 1909

As the most successful monoplane of the era, the Type Onze was produced in a bewildering series of models. A number of these are identified below. See Chapter III for details.

Cross-Channel 1909

Span: 7.8 m
Chord: 2.0 m
Area: 14.0 m² (originally 12.0 m²)
Weight: 300 kg (including pilot and consumables for a two-hour flight)
Engine: 3-cyl. Anzani fan, 30 hp at 1,600 rpm
Speed: 36 mph (according to L'Aérophile; Blériot claimed 42 mph)
Fabric: Continentale rubberized

XI. 1912

The classic Type XI competition machine.
Span: 8.9 m
Chord: 2.3 m
Area: 15.0 m²
Stabilizer Span: 3.1 m
Stabilizer Chord: 1.5 m
Length: 7.8 m
Weight: 300 kg (flight)
Propeller: 2.6 m diameter
Engine: 50-hp Gnôme
Speed: 90 kph
Wheel base: 1.5 m

XI-bis. 1910

The term XI-bis illustrates the problems of Blériot nomenclature. It was first applied to a fantailed machine with a completely covered fuselage that appeared in January 1910 and was abandoned that May. This machine was also incorrectly identified as No. 13. The dimensions of this machine were similar to those of the Type XI, with the exception of the fuselage, which measured 6.6 m.

XI-2. 1912–14

A standard tandem two-place Blériot.
Span: 10.35 m
Area: 19.0 m²
Length: 8.4 m
Height: 2.5 m
Weight: 335 kg (empty)
585 kg (flight)
Engine: 70-hp Gnôme
Gasoline: 90 liters
Oil: 25 liters
Duration: 3 hrs.
Speed: 115 to 120 kph

XI-2. 1913 Hydroaeroplane

Essentially an XI-2 mounted on floats, the hydroaeroplane had a larger wing area than the standard craft in order to carry the additional weight.
Span: 11.05 m
Wing area: 21.0 m²
Elevator area: 3.25 m²
Length: 8.875 m
Engine: 80-hp Gnôme
Speed: 110 kph

XI-2. 1912 Génie

A military version of the XI-2. The Artillerie and other military machines were very similar. See text for discussion of these types.
Span: 9.7 m
Area: 18.0 m²
Length: 8.3 m
Height: 2.5 m
Weight: 320 kg (empty)
550 kg (flight)
Engine: 70-hp Gnôme
Gasoline: 75 liters
Oil: 25 liters
Speed: 110–15 kph

The Génie was designed for easy transport and could be broken down or reassembled in 25 minutes. Crated, the machine measured:
Span: 1.7 m
Length: 7.5 m
Height: 2.25 m

XI-2 bis. 1910

A large, two-place, side-by-side machine unveiled at Reims and Bournemouth in 1910. The craft was flown by Graham-Gilmour as the "Big Bat" Blériot. It was clearly a simple enlargement of the XI-bis.
Span: 11.0 m
Chord: 2.6 m
Length: 8.3 m
Area: 19.0 m²
Area of tailplane and elevator: 34.75 m²
Area of rudder: 3.35 m²
Weight: 350 kg (flight)
Engine: 80-hp Gnôme
Propeller: 2.59 m

XI-3. 1912–13

A tandem three-seat version of the Type XI. The machine saw limited military service. The craft featured triple-tired front wheels.
Span: 11.35 m
Chord: approx. 1.75 m
Area: 25.0 m²
Length: 8.2 m
Weight: 519 kg (flight)
Engine: 140-hp Gnôme

XI. 1912–14 Penguin

The Penguin machines were trainers produced in school and military versions. Dimensions of both versions are given below.

	Military	School
Span:	8.9 m	8.9 m
Area:	15.0 m²	15.0 m²
Length:	7.8 m	7.65 m
Height:	2.5 m	2.4 m
Weight:	265 kg (empty)	220 kg (empty)
	415 kg (flight)	350 kg (flight)
Engine:	50-hp Gnôme	30-hp Anzani
Speed:	95 kph	65 kph

During W.W.I, Blériot Aéronautique, and other firms like Breese, operating under license, produced a severely clipped-wing Penguin, powered by 2-cylinder opposed engines. These craft were strictly ground trainers, incapable of flight.

XI. 1914 Blériot-Gouin (BG) Vision Totale

The final version of the Type Onze was a high-wing two-place monoplane that offered improved visibility and a slightly larger wing area. The aircraft saw service as an observation machine in 1914-15.

Blériot XII, from the rear.

Span: 10.36 m
Chord: 2.15 m
Area: 19.5 m²
Length: 8.427 m
Weight: 600 kg (empty)
Engine: 80-hp Gnôme
Propeller: 2.6 m diameter
Speed: 100 kph

XII. 1909

High-wing monoplane flown in conjunction with Type Onze up to and following the 1909 Reims Meet.
Span: 10.0 m
Area: 22.0 m²
Length 9.5 m
Weight: 600 kg (empty)
Engine: 35–50-hp ENV
Propeller: Chauvière Intégrale, 2.7 m diameter
Speed: 100 kph

XIII. 1910 L'Aérobus

A high-wing canard monoplane designed as a four-place machine but capable of carrying up to eight passengers. It was equipped with ailerons.
Span: 13.0 m
Area: 40.0 m²
Length: 12.0 m
Weight: 600 kg (flight)
Propeller: Chauvière Intégrale, 3.5 m diameter

XIV(?). 1911

There is some evidence to indicate that the "new 50-hp racing monoplane" exhibited at the Paris Salon in December 1911 was identified as No. XIV. The craft featured a single-piece trailing aileron and an all-moving rudder with no vertical stabilizer. In 1913, Blériot mounted a version of this craft on floats; Leo Opdycke has identified it as No. XIV. The craft is similar in appearance to the XI-bis. *Aircraft* for February 1911 identified XIV only as "a new two-passenger machine." The confusion is typical of Blériot nomenclature.
Span: 7.0 m
Area: 14.0 m²
Weight: 226.8 kg
Engine: 50-hp Gnôme
Price: £960
Speed: 128 kph
(Data is for the Paris Salon machine.)

XX. 1911

Flight for February 11, 1911, describes No. XX as being tested at Pau. The wings were adjustable and "arranged to give the smallest surface possible at 100 kph." No further details are offered.

XXI. 1911

Two-place "côté-à-côté" monoplane. See text for details. Lateral control was by wing warping. Some models may have been mounted on floats.

The passenger-carrying Blériot XIII.

Span: 11.0 m
Area: 25.0 m²
Length: 8.24 m
Weight: 370 kg
Engine: 70-hp Gnôme
Propeller: Chauvière Intégrale
Speed: 90 kph
The XXI was also available in a single-seat "Militaire" version.
Span: 8.84 m
Length: 7.0 m
Weight: 249.5 kg
Engine: 50-hp Gnôme

The Blériot XIII.

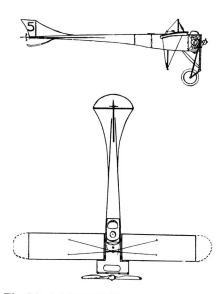

The Blériot XXIII. Drawings by John J. Ide.

XXIII. (sometimes identified as XI-bis)
1911 Type de Concourse

This craft, with its clipped, high-aspect-ratio wings, and sloping cowl, was clearly derived from the shortlived XI-bis of the previous year. It also bears a resemblance to the single-seat Type XII. It debuted at Reims in 1911, where Léon Morane flew it to victory, setting new records for 5, 10, and 20 km. Alfred Leblanc and Gustav Hamel flew even more severely clipped-wing versions in the Gordon Bennett Race at Eastchurch, England, that summer.
Span: 6.71 m (reduced from 7.16 m)
Chord: 3 ft 6 in
Wing area: 9.0 m² (6.75 m² after alteration)
Length: 7.62 m
Engine: 100-hp Gnôme
Propeller: Regy, 7 ft 6 in diameter
Speed: 128 kph

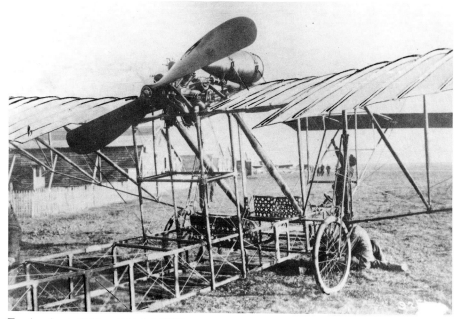

Engine installation, Blériot XIII.

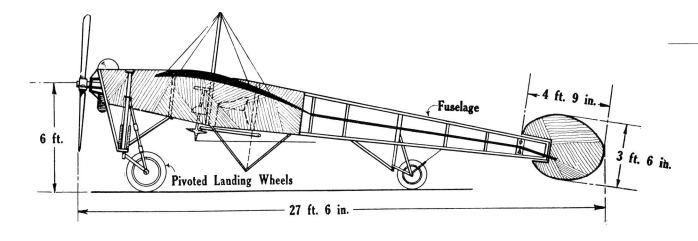

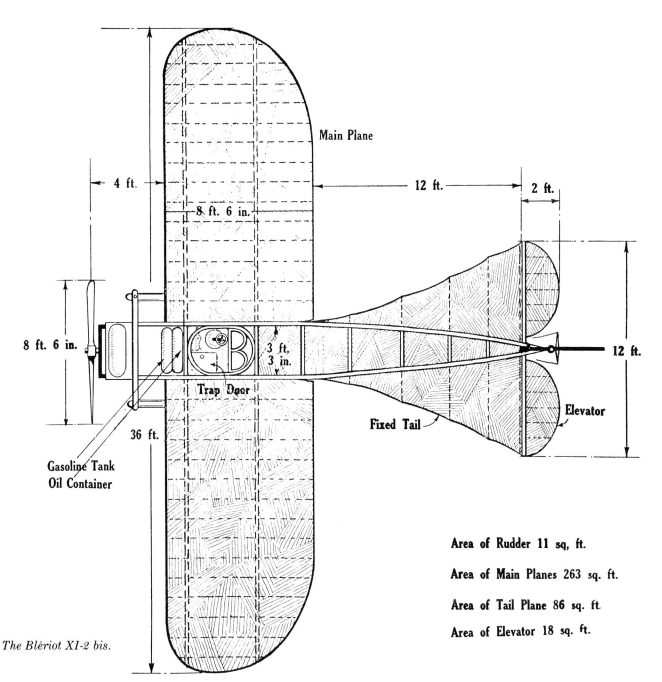

The Blériot XI-2 bis.

Area of Rudder 11 sq. ft.

Area of Main Planes 263 sq. ft.

Area of Tail Plane 86 sq. ft.

Area of Elevator 18 sq. ft.

The Blériot XXIV Aeronef built for E. Archdeacon.

The Blériot XXIV in flight.

XXIV. 1911 Aéronef

A high-wing monoplane multiplace machine derived from the Blériot XIII. The Aeronef was built for E. Archdeacon.

Span: 14.5 m
Area: 40.0 m²
Length: 12.5 m
Weight: 700 kg
Engine: 100-hp Gnôme
Propeller: Normale, 2.85 m diameter

XXV. 1911 Canard

A small canard monoplane, undoubtedly related to the Blériot V of 1907.

Span: 8.9 m
Elevator span: 2.1 m
Length: 5.5 m
Weight: 400 kg

XXVII. 1911

An experimental racing monoplane derived from the Type XI.

Span: 8.9 m
Area: 12.0 m²
Length: 7.5 m
Engine: 70-hp Gnôme
Weight: 430 kg (flight)
Speed: 125 kph

In this case, Blériot departed from his standard practice by mounting the engine only at the rear, abandoning the usual forward mount to save weight.

XXVIII. 1912 Populaire

A small monoplane designed to be produced inexpensively for new pilots. It was produced in both one- and two-place versions.

	One-place	Two-place
Span:	4.8 m	9.75 m
Area:	15.0 m²	20.0 m²
Length:	7.6 m	8.2 m
Weight:	240 kg (empty)	300 kg (empty)
	369 kg (flight)	550 kg (flight)
Engine:	50-hp Gnôme or	70-hp Gnôme
	35-hp Anzani	115 kph
Speed:	100 kph	

XXXIII. 1912 Canard

A large canard related to the Blériot V of 1907 and the XXV of 1911. The XXXIII was a two-place (side-by-side) monoplane.

Span: 10.5 m
Area: 24 m²
Length: 8.0 m
Weight: 330 kg
Engine: 70-hp Gnôme
Propeller: Chauvière
Speed: 115 kph

The Blériot XXVII.

The Blériot XXXIII canard of 1912.

The Blériot XXXIII from the rear.

XXXVI. 1912–13 Blinde

An armored military monoplane exhibited at the Salon de l'Aéro in 1912. Monocoque fuselage construction was employed. A 6-mm skin of paper, sheet cork, and fabric was glued over a form. Aluminum was used in the cowl area.

Span: 12.25 m
Chord: 2.25 m
Area: 25.0 m²
Weight: 375 kg
Engine: 80-hp Gnôme
Speed: 75 mph

40. 1913

A large two-place biplane with a crew nacelle, shown at the Salon de l'Aéronautique in 1913.

Span: 12.7 m
Area: 38.0 m²
Length: 9.15 m
Height: 3.3 m
Weight: 400 kg (empty)
        650 kg (flight)
Engine: 80-hp Gnôme
Speed: 90 kph

42. 1913 Canard

Armored military biplane canard observation craft. Pusher propeller and canard carried on a long forward framework.

Span: 8.9 m
Area: 18.0 m²
Length: 7.3 m
Engine: 80-hp Gnôme

The Blériot XXXVI Blinde (armored) with monocoque fuselage.

The Blériot 42 canard.

43. 1913 Blinde

Two-place military monoplane.
Span: 10.1 m
Area: 10.0 m²
Length: 6.02 m
Weight: 350 kg (empty)
         625 kg (flight)
Engine: 80-hp Gnôme
Speed: 120 kph

45. March 1914

Small, lightweight monoplane reminiscent of the XXI, powered by 6-cyl. Anzani. No details are available.

Blériot-Gnôme 1913

L'Aérophile for April 1913 describes a shoulder-wing monoplane powered by a twin-row 60-hp Gnôme rotary. The aircraft was flown by E. Perreyon to a record altitude of 6,000 m on March 11, 1913. In making the flight, Perreyon wore an oxygen mask, perhaps the first recorded instance of the use of such a device for high-altitude flying.

Special Military 1913

Produced, as the editor of *Jane's* commented, "with considerable secrecy"; no details other than the rough length of the craft (10.0 m) are available to the author. The machine was a shoulder-wing pusher, with the fuselage mounted well above the tail-support frame and undercarriage.

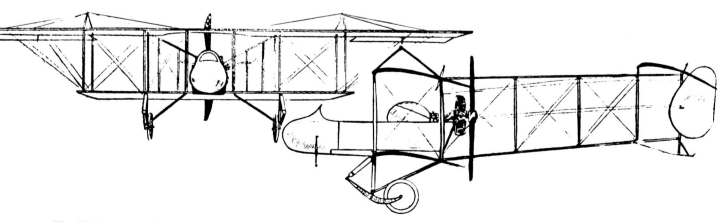

The Blériot 40.

The Blériot 43.

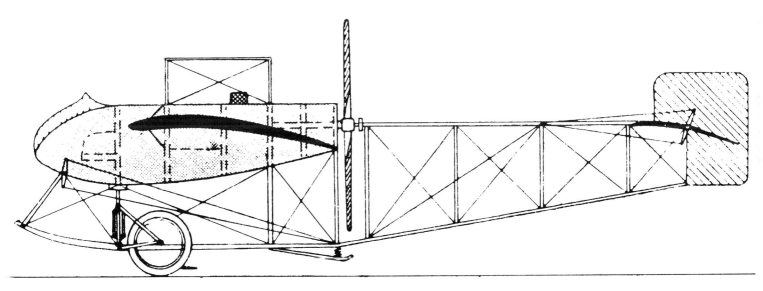

The special military Blériot.

Restored and Reproduction Blériot Aircraft

The following list of restored and reproduction Blériot aircraft is based on information made available through the courtesy of Leonard O. Opdycke and was originally published in *World War I Aeroplanes.*

ORIGINAL BLÉRIOT MACHINES SURVIVING

1. *Argentina:* Museo Nacional de Aeronautica, Aeroparque Airport, Buenos Aires—Used by Teodore Fels in the first crossing of the River Plate.

2. *Australia:* Museum of Applied Arts and Sciences, Sydney—Flown by Maurice Guillaux on the first Melbourne-Sydney airmail flight, 1914. The machine is also reputed to have been used by R. P. Carey to fly the first loop in Australia, 1914.

3. *Belgium:* Musée de l'Armée, Brussels—The wings, seat, and Anzani engine from Jan Olieslagers's Blériot. The Belgian cockade (red center to yellow to black) is incorrect.

4. *Czechoslovakia:* National Technical Museum, Prague—Apparently a Blériot replica built by Kaspar in 1910. It is powered by a 65-hp Daimler-Benz engine.

5. *England:* RAF Museum (on loan from the Royal Aeronautical Society), Hendon—Said to be a 1912 military machine "found in France" by R. G. J. Nash in the 1930s. While the machine is generally thought to be genuine, it is interesting to note that A. R. Weyl later reported that he had constructed a replica Blériot for Nash. The machine is reputed to be #164 and is powered by a 25-hp Anzani. The RAS purchased the aircraft in 1953.

6. *England:* Shuttleworth Trust, Biggleswade—This airplane is one of the original Humber-built Blériot types flown at the Blériot School at Hendon in 1910. Sarah Merchant bought the machine and sold it to A. Grimer, who crashed it late in 1912. R. O. Shuttleworth acquired it in 1935.

7. *England:* Mosquito Museum, St. Albans—Aircraft #225 is powered by a 25-hp Anzani. The museum acquired the machine from Cdr. L. D. Goldsmith, who had stored it at RAF Colerne.

8. *France:* Musée de l'Air, Paris—Apparently a 1909 aircraft, no details.

9. *France:* Musée de l'Air, Paris—XI-2 Militaire, 1913. Wings are marked Pégoud. The aircraft is #686.

10. *France:* Musée de l'Air, Paris—As above, #878.

11. *France:* Musée des Techniques CNAM, Paris—This is the original cross-Channel Blériot.

12. *Germany:* Deutsches Museum, Munich—Apparently a 1909 aircraft rebuilt from Rozendaal drawings. Presented by Dr. Paul Gens of Bavaria on February 25, 1912.

13. *Italy:* Museo del Volo (Aeronautica Militare Italiana), Turin—An STI-built XI-bis, 1913.

14. *Italy:* Museo dell'Instituto Storico, Rome—Flown by Carlo Piazza. Blériot flew this machine on the twentieth anniversary of the Channel flight.

15. *New Zealand:* A Blériot aircraft for which no details are available is reportedly located in Omakau, N.Z.

16. *Norway:* Norsk Teknisk Museum, Oslo—The aircraft flown by Lt. Trygve Gran on the first aerial crossing of the North Sea, July 30, 1914, from Crudden Bay, Scotland, to Reutangen. The aircraft is #794.

17. *Sweden:* Tekniska Museet, Stockholm—Presumably a Blériot XI, no details are available.

18. *Switzerland:* Transport Museum, Lucerne—A 1913 aircraft of the type flown by Oscar Bider over the Pyrenees. This is #23 and is powered by a 70-hp Gnôme.

19. *USA:* Old Rhinebeck Aerodrome, Rhinebeck, New York—A 1911 American Aeroplane Supply House machine with a Gnôme engine. This aircraft was obtained from the U.S. Marine Corps Museum in 1976.

20. *USA:* Old Rhinebeck Aerodrome, Rhinebeck, New York—Cole Palen found the remains of Blériot #56, which had flown and crashed at the Harvard Air Meet. The wreckage was in the hands of William Champlin from 1910 to 1956.

21. *USA:* Old Rhinebeck Aerodrome, Rhinebeck, New York—This is the first Blériot to come to the USA, Rodman Wannamker's #153, later flown by the Bergdolls. The original Anzani is now exhibited in the Franklin Institute, Philadelphia.

22. *USA:* Henry Ford Museum, Dearborn, Michigan—This Blériot, powered by a 50-hp Gnôme, was acquired from Edward Costello. The fabric is gray-green.

23. *USA:* Joseph Kepleman—No details.

24. *USA:* D. White—No details.

25. *USA:* Iowa Department of History and Archives, Des Moines—One of six Blériot XIs brought to the USA by John Moisant in 1910. Following a crash it was shipped to Boston, then to Santa Paula, California. Harry Prevolt acquired and restored the aircraft, using 50 percent of the original fuselage, then sold the machine to E. D. Weeks in May 1961. Weeks presented the craft to the state of Iowa in 1966. Provolt rented the aircraft to Frank Tallman for use in the film *Lafayette Escadrille*.

REPRODUCTION AIRCRAFT

1. *Canada:* National Aernautical Collection, Rockcliffe—This machine was constructed by Jack Hamilton of Palo Alto, California, from Blériot plans, in 1911. It features a Gibson propeller and parts obtained from the California Aero Manufacturing Supply House in San Francisco. It was purchased by James Mathiesen and James Nissen in 1953 and rebuilt by James Moser. It

was sold to the National Aeronautical Collection on December 19, 1971. The airplane is powered by a 3-cylinder Elbridge engine, #301, which dates to 1911.

2. *England:* RAF Museum, Hendon—Apparently built from spare and new parts, this airplane is powered by an Anzani.

3. *England:* Salisbury Hall—Same as above.

4. *England:* Midland Aircraft Preservation Society, Coventry—This aircraft was built by RAF apprentices for a 1959 RAF Royal Tournament. It was restored in 1971 by the MAPS as a Humber aircraft with a Humber engine.

5. *England:* Science Museum, London—Both airplane and engine were constructed by J. A. Prestwick for H. J. Harding in 1910. The machine features square wingtips and ailerons and is powered by a 45-hp JAP V-8. It first flew on April 10, 1910.

6. *England:* Mike Beach, Twickenham—This reproduction was reported under construction in 1978. No further details are available.

7. *France:* Jean Salis, La Ferte Alais—Salis has flown the Channel with this craft powered by a 3-cylinder Anzani. A 1925 Potez is now fitted to the machine. Some original parts were used in its construction. Originally flown under French registration F-WERV, it is now F-PERV.

8. *Germany:* Hermann Ring, Phillipsburgerstrasse 8, Romerburg II—Work began on this machine in January 1976. It is powered by a Rossel-Peugeot rotary.

9. *Italy:* Museo Nazionale della Scientifico del Volo, Milan—Either a museum- or Macchi-built Type XI of the sort flown during the war in Tripoli.

10. *Italy:* Centro Storico Scientifico del Volo, Turin—A reproduction of the Type XI-2 flown by Italian pilots in Tripoli.

11. *Peru:* Air Force Academy, Los Palmas—Reproduction Type XI.

12. *Poland:* Polish Aviation and Astronautic Museum, Krakow—A Type XI built by Pawel Zolotow of Lublin based on drawings published in *Modellarz*. The aircraft is powered by a Salmson 9AD (#105090) with a Piper J-3 propeller.

13. *Sweden:* National Museum of Science and Technology, Stockholm—This Blériot copy was constructed by Enoch Thulin for use at the Thulin aviation school, 1916–18. The museum acquired the machine in 1928 and restored it 1936–38. The engine is a 90-hp Thulin. (Museum #TK-5868.)

14. *Sweden:* National Museum of Science and Technology, Stockholm—Hjalmar Nyrop built this Blériot copy in his boatyard at Landskorne in the fall of 1911. It was purchased by E. O. Neumuller and given to the Swedish Navy the same year. The museum acquired the aircraft and engine, a 50-hp Gnôme, in 1936.

15. *USA:* USAF Museum, Wright-Paterson AFB—Ohio earlybird Ernest Hall built this 1911 Blériot on the basis of drawings and photos. It is powered by a 3-cylinder Anzani. The aircraft was recovered in 1974.

16. *USA:* Morgan Hill, California—Stan Hill build this reproduction Type XI at the Calgary Institute of Technology in the early 1950s. It has made

one flight across the Channel and was used in the filming of *Lafayette Escadrille*. It is powered by a 65-hp Continental.

17. *USA:* Owl's Head Transportation Museum, Owl's Head, Maine—This machine was constructed by Harry Prevolt and was owned by Wings and Wheels before coming to Owl's Head in 1976. It is registered #N7899C.

18. *USA:* A. H. Lane, Brownsville, California—This is an XI-2 constructed by Alexander Block in 1910–11. Powered by an 85-hp Continental, it is registered #N3433.

19. *USA:* Wings and Wheels, Orlando, Florida—This Type XI was constructed by Walter Bullock and exhibited at the Johan Larson Museum, Minneapolis, before being acquired by Wings and Wheels.

20. *USA:* Tallmantz Aviation, Santa Anna, California—Built by Revis D. Henry, this Type XI was first flown on March 10, 1968. It is powered by a 65-hp Continental and features ailerons. Tallmantz purchased the craft in 1970.

21. *USA:* Connecticut Aviation Historical Society, Bradley Field, Hartford—This is another Ernie Hall machine (see #15 above). Built in 1910–11, it has been heavily restored, so that the extent of the original material remaining is difficult to determine. The aircraft was heavily damaged in the tornado that devastated the Bradley Field Museum in 1979.

MISCELLANEOUS BLÉRIOT AIRCRAFT

1. *Canada:* National Aeronautical Collection, Rockcliffe—A Blériot Type XI constructed by Robert McDowell of Owen Sound in 1912. It passed through the hands of a number of Canadian pilots before being acquired by the NAC.

2. *England:* RAF Museum, Hendon—A Blériot XXVII built for Alfred Leblanc. It is powered by a 50-hp Gnôme and is factory #433.

3. *USA:* Bettis Field, Pittsburgh, Pennsylvania—This is one of many Type XIs which have disappeared in recent years. In this case, the machine was a standard feature of Pittsburgh art shows prior to WW II. The air field no longer exists, and no details as to the fate of the aircraft are available.

Drawings
of the Blériot XI

The finest set of contemporary drawings of the Blériot XI were prepared by the Dutch engineer John Rozendaal for the German auto-aero journal *Der Motorwagen* (Jahrg. 15, Hefte 1, 10 Jan.–20 Feb. 1912, pp. 12–14; 107–110). They are reproduced here through the courtesy of Leonard E. Opdycke and *World War I Aeroplanes*.

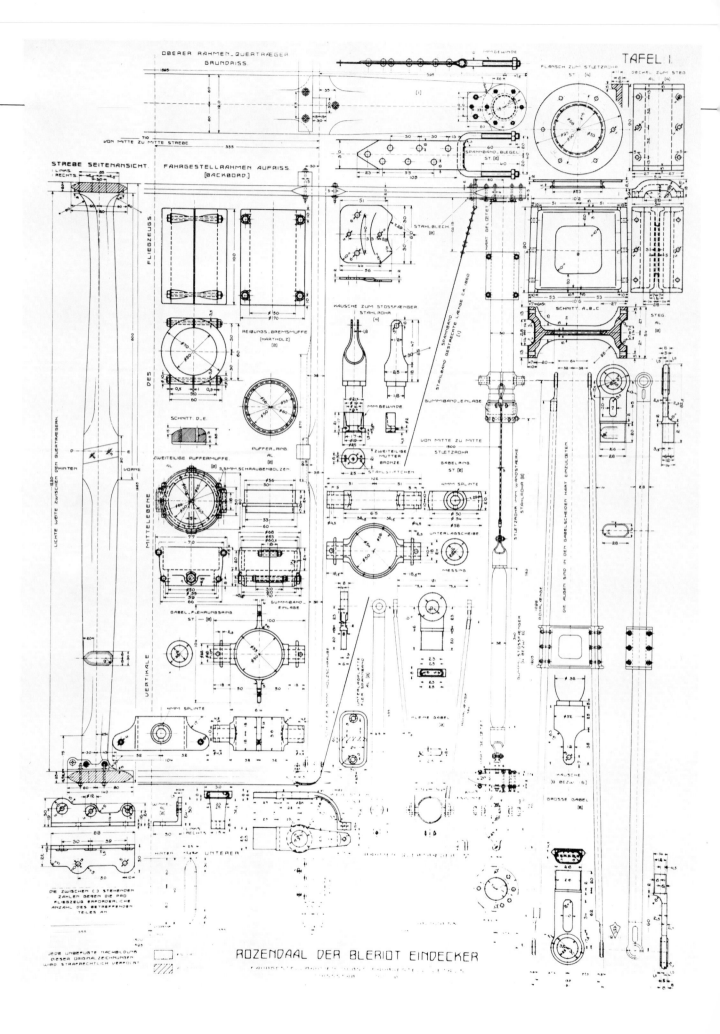

TAFEL I.

ROZENDAAL DER BLERIOT EINDECKER

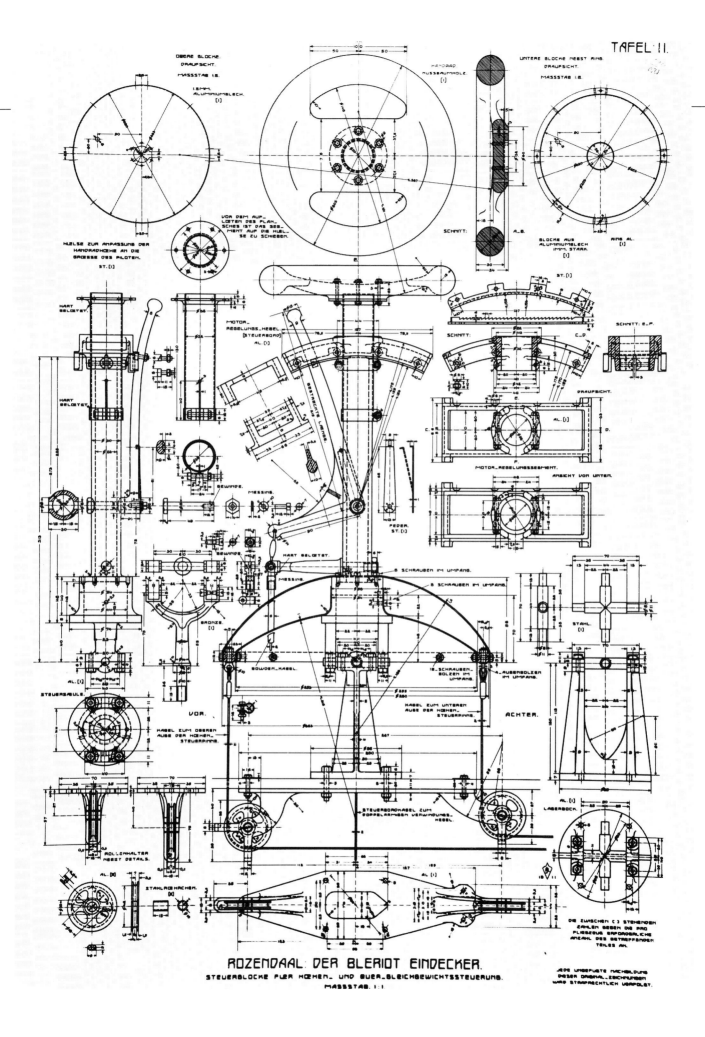

TAFEL II.

ROZENDAAL: DER BLERIOT EINDECKER.
STEUERBLOCKE FUER HOEHEN- UND QUER-GLEICHGEWICHTSSTEUERUNG.
MASSSTAB 1:1

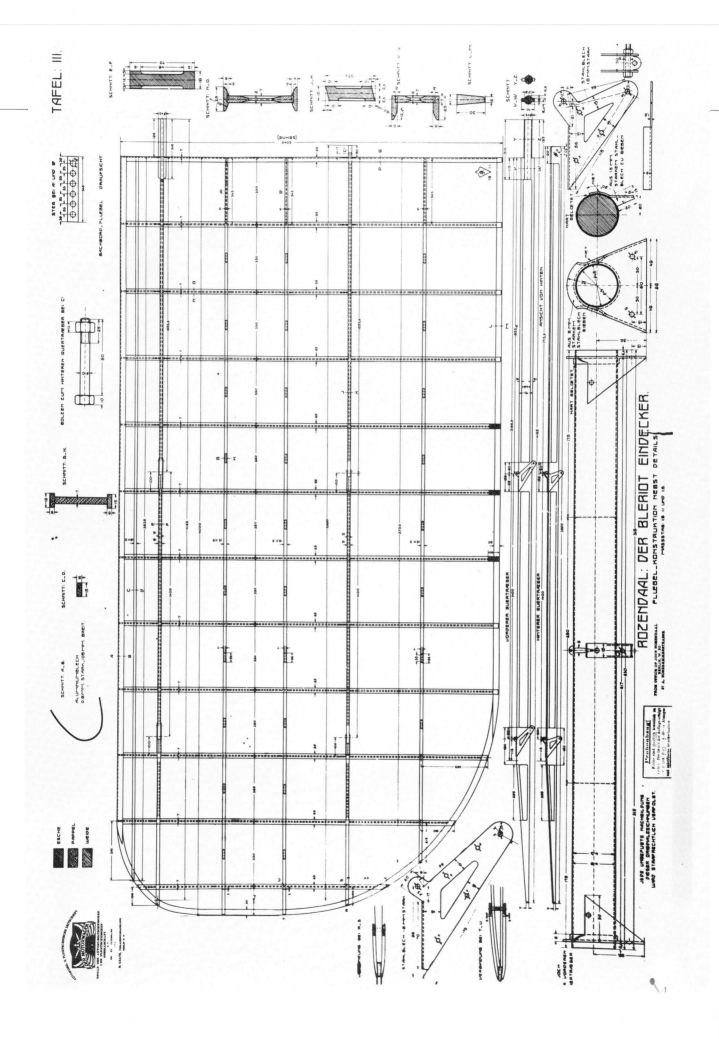

ROZENDAAL: DER BLÉRIOT EINDECKER.
SCHWANZ_TRAGDECK HOEHENSTEUER_FLOSSEN NEBST DETAILS.

ROZENDAAL: DER BLERIOT EINDECKER.

KONSTRUKTION DES RUMPFES

MASSSTAB 1:10

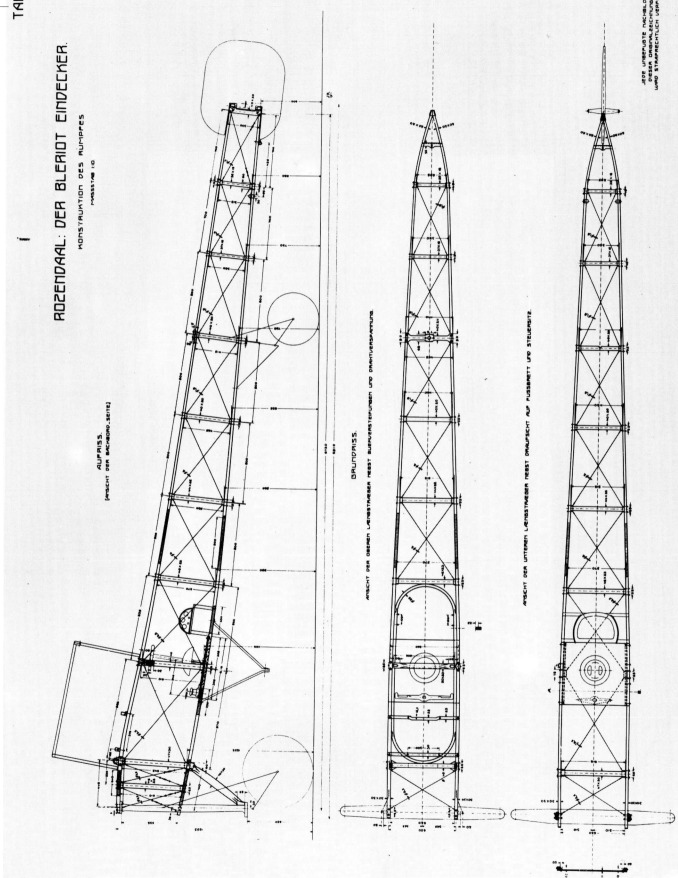

AUFRISS.

(ANSICHT DER BACKBORD-SEITE)

GRUNDRISS.

ANSICHT DER OBEREN LÄNGSTRAEGER NEBST QUERVERSTEIFUNGEN UND DRAHTVERSPANNUNG.

ANSICHT DER UNTEREN LÄNGSTRAEGER NEBST DRAUFSICHT AUF FUSSBRETT UND STEUERSITZ.

SCHNITT A-B

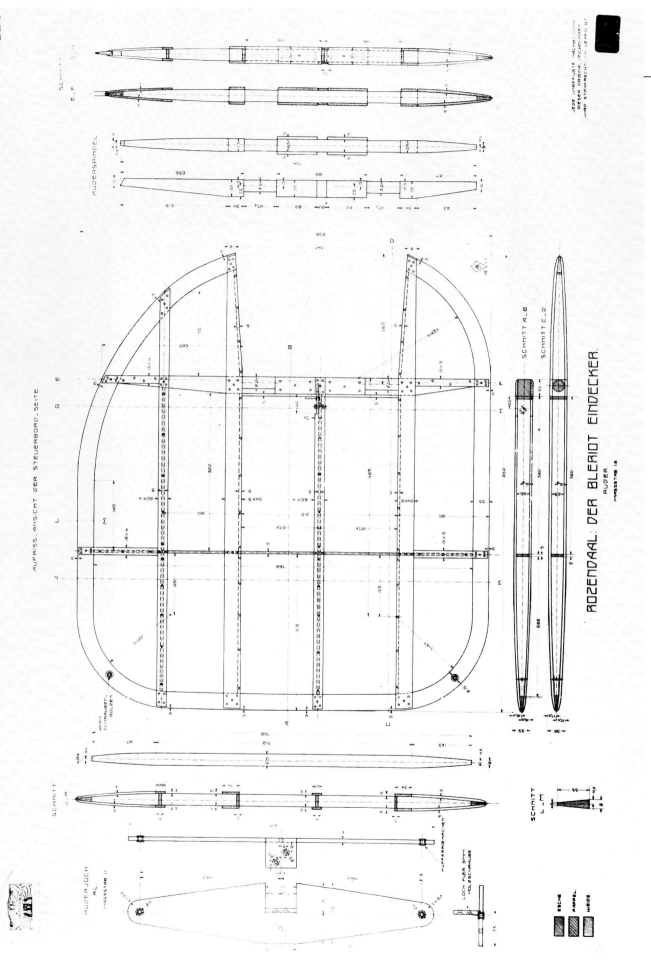

ROZENDAAL: DER BLERIOT EINDECKER.
RUDER

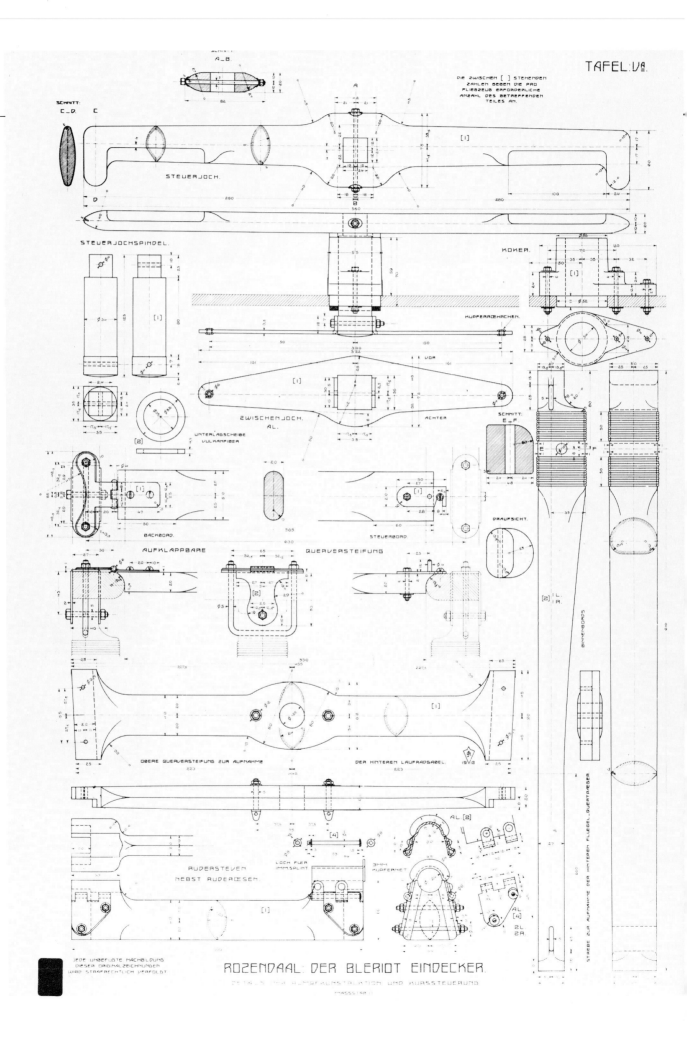

TAFEL: Vᵃ.

ROZENDAAL: DER BLERIOT EINDECKER.

Notes

In preparing this volume the author combed through countemporary aeronautical journals for scraps of information on Blériot aircraft. To have provided a citation for every bit of information would have broken up the text unnecessarily and created an unwieldy notes section. The author has chosen, therefore, to cite only direct quotations or pieces of controversial data. The reader interested in uncovering the source of some bit of technical information given but not cited in the text can only be directed to Paul Brockett's fine *Bibliography of Aeronautics* (Washington D.C., 1909–14) and to the indexes in individual volumes of *Flight*, *L'Aérophile*, and various other journals.

Introduction

1. Claude Graham-White and Harry Harper, *The Aeroplane: Past, Present, and Future* (London, 1911), pp. 44–49. Other data on the number of aircraft built and flown was drawn from Blériot sales brochures, i.e., *L. Blériot Recherches Aéronautiques* (Paris, 1911); *Blériot Aéronautique, 1912; Les Usines Blériot;* and *Les Monoplanes Blériot.*

2. "Twenty-Five Miles Across Country—Blériot's Great Flight," *Flight*, 17 July 1909, p. 423.

Chapter I: "Un Aviateur Militant"

1. "The First Paris Aeronautical Salon," *Flight*, 2 Jan. 1909, p. 6. See also, "Au Premier Salon de l'Aéronautique," *L'Aérophile*, 15 Jan. 1909, p. 42; "An Aeronautic Salon," *New York Tribune*, 10 Jan. 1909; "Wright Aeroplane Caught Crowds," *New York World*, 3 Jan. 1909; "Worship of Airship," *Washington Post*, 27 Dec. 1909; "Au Salon de l'Aéronautique," *Paris Press*, 28 Dec. 1908.

2. Biographical material on Blériot has been drawn from a variety of sources. Some of the best are: Simone Rubel Blériot, "Souvenirs d'Enfance," *Icare* (1979), p. 97; Michel Lhospice, *Match Pour La Manche* (Paris, 1964); Charles Fontaine, *Comment Blériot A Traversé La Manche* (Paris, 1909).

3. "En Vol Plane Au-dessus de la Seine," *L'Aérophile*, July 1905, p. 161.

4. Gabriel Voisin, *Men, Women and 10,000 Kites* (London, 1963), pp. 139–43; see also "En Vol Plane" (note 3).

5. Ibid.

6. Ibid. "L'Aéroplane Blériot," *L'Aérophile*, June 1906, pp. 148–49.

7. Voisin, *Kites* (note 4).

8. "L'Aéroplane Blériot," *L'Aérophile*, Oct. 1906, p. 250; "Les Aéroplanes Blériot et Voisin," *L'Aérophile*, Dec. 1906, p. 295.

9. Ibid.

10. "First Paris Aeronautical Salon," *Flight*, 2 Jan. 1909, p. 6.

11. Voisin, *Kites* (note 4).

12. Ferdinand Collin, *Parmi les Précurseurs du Ciel* (Paris, 1943), p. 35.

13. Ibid.

14. "De la Rapidité avec laquelle les Aviateurs s'Orientent Vers l'Avenir," *L'Aérophile*, Feb. 1907, p. 30; "L'Aéroplane Louis Blériot," *L'Aérophile*, Apr. 1907, p. 96; "Nouveaux Essais de l'Aéroplane Louis Blériot," *L'Aérophile*, May 1907, p. 126.

15. For information on the Blériot/Esnault-Pelterie patent dispute, see Lhospice (note 2).

16. "Essais Heureux du Nouvel Aéroplane Blériot," *L'Aérophile*, July 1907, p. 194; "L'Aéroplane Blériot," *L'Aérophile*, Aug. 1907, p. 230; "Expérience de L'Aéroplane Blériot, à

Issy-les-Moulineaux, le 15 Juillet 1907," *La Revue de l'Aviation*, 15 July 1907.

17. Louis Blériot and Edward Ramond, *La Gloire des Ailes* (Paris, 1927), p. 26.

18. Henry S. Villard, *Contact: The Story of the Early Birds* (New York, 1968), p. 47.

19. "The New Blériot Aeroplane," *Bulletino della Società Aeronautica Italiana*, Aug. 1901 (translation in NASM Library).

20. "M. Blériot franchit 184 mètres," *L'Aérophile*, Sept. 1907, p. 262; "Le Nouvel Aéroplane Blériot de 50 Chevaux," *L'Aérophile*, Nov. 1907, p. 318; "L'Aéroplane Blériot," *L'Aérophile*, 1 Jan. 1908, pp. 9–10.

21. "Les Expériences de M. Blériot," *L'Aérophile*, 15 July–1 Aug. 1908, p. 308; "Le Monoplan 'Blériot VIII-ter,' " *L'Aérophile*, 15 Sept. 1908, p. 360; "Les Expériences de Louis Blériot," *L'Aérophile*, 1 Oct. 1908, p. 387; "A Issy-les-Moulineaux," *L'Aérophile*, 15 Oct. 1980, p. 410; "Blériot et le Prix de la hauteur," *L'Aérophile*, 1 Nov. 1908, p. 434; "Les Expériences de Louis Blériot," *L'Aérophile*, 15 Nov. 1908, p. 460.

22. Alphonse Berget, *The Conquest of the Air* (New York, 1909), p. 252.

23. Blériot, *La Gloire* (note 17), p. 97.

Chapter II: "C'est un Triomphe!"
1. Charles Dollfus, "J'ai Rencontré Blériot pour la Première Fois le 23 Juin 1907," *Icare* (1979), p. 89. My thanks to C. H. Gibbs-Smith for his thoughts on the design controversy.

2. C. H. Gibbs-Smith, *The Rebirth of European Aviation, 1902–1908* (London, 1974), p. 286.

3. Ross Browne, interview with Kenneth Leish, Columbia University Oral History Collection, transcript, p. 17.

4. All information on the early flights of the Blériot XI has been drawn from scattered references in *L'Aérophile, Flight,* and *L'Aéronaut.*

5. Ibid.

6. Ferdinand Collin, *Parmi les Précurseurs du Ciel* (Paris, 1943), p. 50.

7. "Monetary Encouragement of the Industry in France," *Flight*, 3 July 1909, p. 388.

8. "Prizes for Flight," *Flight*, 17 Apr. 1909, p. 216.

9. "Channel Flight," *London Daily Mail*, 20 July 1909.

10. "Continental News," *The Aero*, 27 July 1909, p. 156; "Wilbur Wright on the Channel Flight," *Flight*, 24 July 1909, p. 441.

11. "Channel Flight," *London Daily Mail*, 22 July 1909.

12. The author has traced the story of the Channel "race" through the pages of *Flight, The Aero, L'Aérophile,* and a collection of news cuttings in the A. G. Bell scrapbooks, NASM.

13. Collin, *Précurseurs* (note 6), p. 65.

14. Ibid., 68–69.

15. Louis Blériot and Edward Ramond, *La Gloire des Ailes* (Paris, 1927), pp. 97–98.

16. Blériot's own accounts of his most famous flight are inconsistent in some details. In addition, the accounts of others who were present for takeoff or landing are often in disagreement as to times, distances, and the like. As a result a mythology has grown up around the event. Harry Harper, for example, spread the story that Blériot's engine overheated during the flight and was saved only when Blériot flew through a passing rain shower. In fact, there was no shower, and if there had been the rain would have been of little value in cooling the engine.

Blériot himself, in one news account, reported that he neglected to operate the oil pump at one point during the flight, causing the engine to cough and die. In other accounts, he failed to mention this fact.

The account given here has been pieced together from several of Blériot's descriptions of the flight. The author has melded those elements of the various accounts that seem most consistent with the facts. The accounts used were: Blériot, *La Gloire* (note 15), pp. 99–103; "Blériot: First Man to Fly the Channel," *London Daily Mail* (overseas edition), 13 July 1909; Charles Fontaine, *Comment Blériot A Traversé La Manche* (Paris, 1909), chaps. 2–5; "Blériot Tells of His Flight," *New York Times*, 26 July 1909; "Blériot Tells of Great Feat," *New York Herald*, 26 July 1909. See

also "La Traverse de la Manche," *La Revue Aérienne*, 10 Aug. 1909; "The First Cross-Channel Flight," *Aeronautics*, Aug. 1909; and appropriate issues of *Flight* and *L'Aérophile* for contemporary comment.

17. "Blériot: First Man to Fly the English Channel," *London Daily Mail* (overseas edition), 31 July 1909.

18. Ibid.

19. "Editorial Notes," *The Aero*, 3 Aug. 1909, p. 162.

20. Quoted in C. H. Gibbs-Smith, "The Man Who Came by Air," *Shell Aviation News*, June 1959, p. 2.

21. Ibid., p. 6.

22. *Pall Mall Gazette*, 27 July 1909.

23. *London Daily Mail*, 26 July 1909.

24. H. G. Wells quoted in Gibbs-Smith, "The Man" (note 20), p. 7.

Chapter III: "Type Onze"
1. It is difficult to understand where the belief that Blériot abandoned flying after the Istanbul crash arose. A perusal of *Flight* and *L'Aérophile* through 1912 reveals repeated Blériot flights long after the crash.

2. Blériot sales brochures and scattered production figures in contemporary aero journals are the only sources of information on the growth of Blériot Aéronautique. Contacts with the Musée de l'Air failed to produce any company production or financial records. See "The Commercial Side," *Flight*, 31 July 1909, p. 406, for example.

3. Jean Conneau, *My Three Big Flights* (New York, 1912), p. 8.

4. Earle Ovington, "Diary," in Adelaide Ovington, *An Aviator's Wife* (New York, 1920), pp. 20–21.

5. C. G. Grey, "The Gordon Bennett Race," *The Aeroplane*, 6 July 1912, p. 102.

6. Henry Woodhouse, "The Gordon Bennett Cup," *Flying*, Sept. 1912, p. 6.

7. Ovington, *Aviator's Wife* (note 4), p. 126.

8. Ibid., p. 129. Additional information on Queen and AASH monoplanes has been drawn from: C. M. Adrich, "Millionaire an Aviator," *Fly*, Jan. 1910; Walter E. Hapgood, "Aeronautic Show

a Success," *Fly*, Mar. 1910; "American Made Monoplanes," *Fly*, May 1912, p. 25; "The Queen Monoplane," *Aeronautics*, Oct. 1911, p. 125; Grover Loening, *Our Wings Grow Faster* (New York, 1935), pp. 21–25; and various Queen and AASH sales brochures.

Chapter IV: The Problem with Monoplanes

1. "The Fatal Accident to M. Léon Delagrange," *Flight*, 8 Jan. 1910, p. 30.

2. "Delagrange Accident," *Flight*, 8 Jan. 1910, p. 31.

3. Ibid.

4. There is no single source for total aeronautical fatalities prior to 1913. The author's figures have been compiled from the following partial listings, filled in with reference to the "Necrology" column in *L'Aérophile* and notes in *Flight* and *The Aeroplane*: Claude Grahame-White and Harry Harper, *The Aeroplane: Past, Present, and Future* (London, 1911), p. 110; "Notes," *The Aeronautical Journal*, Jan. 1911, p. 27; "Notes," *The Aeronautical Journal*, Oct. 1911, p. 147; "Aeroplane Fatalities," *The Aeronautical Journal*, July 1912, p. 211; "Fatalities," *The Aeronautical Journal*, Oct. 1912, p. 256; "The Uses of Accidents," *Flight*, 28 Jan. 1911, p. 69.

5. Granville E. Bradshaw, "Aeroplanes as Mechanical Constructions," *Flight*, 29 June 1912, p. 593.

6. John H. Ledeboer, "The Study of Dynamic Flight," *Aeronautics*, Dec. 1910, pp. 179–81.

7. R. F. Macfie, "Head Resistance: Has It Been the Cause of Monoplane Disasters?" *Flight*, 9 July 1910, p. 523.

8. Louis Blériot, "Monoplane Failures: M. Blériot's Discoveries and Report to the French Government," *Flight*, 30 Mar. 1912, p. 284.

9. Ibid. For other opinion and reaction to the monoplane problems, see: "The Stresses On Monoplane Wings," *Engineering*, 12 Jan. 1912, p. 53; "The Strength of Monoplane Wings," *Engineering*, 15 Dec. 1911, p. 806; "Monoplane Wing Failures," *Engineering*, 15 Mar. 1912, p. 364; "Accidents to Monoplanes," *Engineering*, 7 Feb. 1913, p. 201.

10. "Head Resistance and Wing Stresses," *Flight*, 13 Apr. 1912, p. 327. See also "Negative Pressure on Wings," *Flight*, 28 Sept. 1912, p. 873.

11. "Head Resistance and Wing Stresses," *Flight*, 6 Apr. 1912, p. 303.

12. "Resistance Tests with Blériot Machines," *Flight*, 15 June 1912, p. 543.

13. C. G. Grey, "In Memoriam: Douglas Graham Gilmour," *The Aeroplane*, 22 Feb. 1922, p. 182.

14. "Matters of Moment," *The Aeroplane*, 29 Feb. 1912, p. 195.

15. "The Ban on Monoplanes," *Flight*, 21 Sept. 1912, p. 848. See also "The War Office and the Monoplane," *Flight*, 28 Sept. 1912; "The Ban on the Monoplane," *The Aero*, 12 Oct. 1912, p. 248.

16. "Monoplanes: The Official Report of the Departmental Committee on the Accidents to Monoplanes," *The Aero*, Apr. 1913, pp. 122–25.

17. Griffith Brewer, "The Collapse of Monoplane Wings," *Flight*, 11 Jan. 1913, p. 33.

18. P. James, "Notes Sur les Accidents d'Aéroplanes Provenant de la Rupture des Ailes," *L'Aérophile*, 15 July 1917, pp. 341–44; idem., "Notes Sur le Calcul et la Construction des Ailes de Monoplan," *L'Aérophile*, 15 Sept. 1911, pp. 428–32.

19. The author owes a genuine debt to Dr. Howard Wolko for his investigation of both wing torsional divergence and beam column failure. Gen. B. Kelsey, NASM Lindbergh Professor, has also been extraordinarily helpful during the course of the investigation.

20. "Pégoud," *Flight*, 27 Sept. 1913, p. 986.

21. "M. Pégoud's Blériot Experiments," *Flight*, Oct. 1913, p. 1,008.

22. "Pégoud and his Flight," *Flight*, 4 Oct. 1913, p. 1,086.

23. "More Exhibitions by Pégoud," *Flight*, 18 Oct. 1913, p. 1,153. See also "Pégoud: Upside Down Flying at Brooklands," *Flight*, 27 Sept. 1913; Louis Blériot, "Apropos of Looping-the-loop," *Flying*, Oct. 1913; A. Pégoud, "Why I Looped the Loop," *Flight*, Oct. 1913, p. 30; "Flying Upside Down," *Literary Digest*, 28 Oct. 1913, p. 749.

24. C. G. Grey, "Louis Blériot," *The Aeroplane*, 12 Aug. 1936, p. 211.

Chapter V: Restoring the Domenjoz Blériot

1. Most biographical information on John Domenjoz has been drawn from the biographical files of the NASM Library. Harold E. Moorehouse's unpublished biographical sketch of Domenjoz (author's file) was also useful.

2. "John Domenjoz," *Chirp*, Jan. 1938, p. 3.

3. Earle Ovington, "The Gnôme Rotary Engine: The Airmen's Chief Reliance," *Scientific American*, 14 Sept. 1912, pp. 218–19.

4. Ibid

Appendix B: Flying the Blériot XI

1. Frank Tallman, *Flying the Old Planes* (Garden City, N.Y., 1973), p. 26.

2. Earle Ovington, "How to Operate a Blériot Monoplane," *Aero*, 6 Apr. 1912.

3. Tallman, *Old Planes* (note 1).

4. Ovington, "Blériot Monoplane" (note 2).

5. John W. Underwood, *Aerobats in the Sky* (Glendale, Cal., 1972).

6. W. H. Phipps, "The Danger of the Lifting Tail and Its Probable Bearing on the Death of Miss Quimby," *Aircraft*, Aug. 1912, p. 182.

7. Ross Browne, interview with Kenneth Leish, Columbia University Oral History Collection, transcript, p. 17.

8. Allen Wheeler, *Building Aeroplanes for Those Magnificent Men* (London, 1965), p. 72.

Appendix C: Select Blériot Aircraft, 1901–1914

1. "Blériot Monoplanes," *Aeronautics*, May 1911, p. 34.

Selected
Bibliography

While the notes serve as a guide to some of the major books and articles consulted during research on this volume, they are by no means a complete bibliography. The following volumes were also useful as general surveys of the field, although they were not cited.

All The World's Aircraft. London, 1909–19.

Dollfus, Charles, and H. Bouché. *Histoire de l'Aéronautique.* Paris, 1932.

Été né, Albert. *Avant les Concordes.* Paris, 1961.

———. *La Victoire des Concordes.* Paris, 1970.

Ferris, Richard. *How It Flies.* New York, 1910.

Gibbs-Smith, C. H. *Aviation: An Historical Survey From Its Origins to the End of World War II.* London, 1970.

———. *A Directory and Nomenclature of the First Aeroplanes.* London, 1966.

———. *The Inventors of the Aeroplane.* London, 1965.

Hayward, Charles B. *Building and Flying an Aeroplane.* Chicago, 1912.

Lhospice, Michel. *Match Pour la Manche.* Paris, 1964.

Loening, Grover Cleveland. *Monoplanes and Biplanes.* New York, 1911.

Loughead, Victor. *Vehicles of the Air.* Chicago, 1911.

Mortane, Jacques. *Les Héros de l'Air.* Paris, 1930.

Sauvage, Roger. *Les Conquerants du Ciel.* Paris, 1960.

Vivien, F. Louis. *Description Detailée du Monoplane Blériot.* Paris, 1911.

Weeks, E. D. "Blériot Models II to XII." *American Aviation Historical Society Journal*, Winter 1964, pp. 288,293.

———. "Blériots—Past and Present." *American Aviation Historical Society Journal*, Summer 1963, pp. 100–110.

World War I Aeroplanes, May 1975 (Blériot issue).